MULTIOBJECTIVE OPTIMIZATION METHODOLOGY

A JUMPING GENE APPROACH

INDUSTRIAL ELECTRONICS SERIES

Series Editors:
Bogdan M. Wilamowski
J. David Irwin

PUBLISHED TITLES

Multiobjective Optimization Methodology: A Jumping Gene Approach
K.S. Tang, T.M. Chan, R.J. Yin, and K.F. Man,
City University of Hong Kong

The Industrial Information Technology Handbook
Richard Zurawski, ISA Corporation

The Power Electronics Handbook
Timothy L. Skvarenina, Purdue University

Supervised and Unsupervised Pattern Recognition: Feature Extraction and
Computational Intelligence
Evangelia Micheli-Tzanakou, Rutgers University

Switched Reluctance Motor Drives: Modeling, Simulation, Analysis, Design, and
Applications
R. Krishnan, Virginia Tech

MULTIOBJECTIVE OPTIMIZATION METHODOLOGY

A JUMPING GENE APPROACH

K.S. Tang • T.M. Chan • R.J. Yin • K.F. Man

CRC Press
Taylor & Francis Group
Boca Raton London New York

CRC Press is an imprint of the
Taylor & Francis Group, an **informa** business

CRC Press
Taylor & Francis Group
6000 Broken Sound Parkway NW, Suite 300
Boca Raton, FL 33487-2742

First issued in paperback 2017

ISBN-13: 978-1-4398-9919-9 (hbk)
ISBN-13: 978-1-138-07255-8 (pbk)

Library of Congress Cataloging-in-Publication Data

Multiobjective optimization methodology : a jumping gene approach / K.S. Tang ... [et al.].
    p. cm. -- (Industrial electronics series)
  Includes bibliographical references and index.
  ISBN 978-1-4398-9919-9 (hardback)
  1. Multiple criteria decision making. 2. Mathematical optimization. 3. Genetic algorithms. 4. Biotechnology--Mathematical models. I. Tang, K. S., 1967-

QA402.5.M8586 2012
519.6--dc23

                 2011045647

Visit the Taylor & Francis Web site at
http://www.taylorandfrancis.com

and the CRC Press Web site at
http://www.crcpress.com

Contents

Preface

The phenomenon of jumping genes was initially discovered by Nobel Laureate Barbara McClintock in her work on corn plants in the 1940s. It was found that some mobile genetic elements could transpose (jump) from one position to another position in a horizontal fashion within the same chromosome or even to other chromosomes. This finding was distinctive as it differed from the classical concept of gene transmission, which is conceptually performed from generation to generation in a vertical sense.

Although research work on jumping genes is generally active in bioscience and biomedical fields, there is little effect on and study in other scientific areas. What was intriguing in the first place was the concept of exploring the mechanism of jumping genes in evolutionary computation. It was a wild guess at the beginning. Then, the guess was not unfounded. Given the formulation of a genetic algorithm, which is mainly subjected to selection, crossover, and mutation as the vital components for gene manipulations in vertical transmission, an evolutionary process that may also comprise horizontal transmission is therefore nothing more than a logical assumption. This is how the jumping gene was born and comes to play its significant part in furnishing the unique evolutionary cycle.

Despite being the first of its kind to introduce the topic of jumping genes outside bioscience/medical areas, this book nonetheless stands solemnly on evolutionary computational ground. It therefore unequivocally requires substantial engineering insight and endeavors for ironing out a number of practicality issues so that the essence of the jumping gene algorithm can be brought out convincingly as well as scientifically. It has to show its robustness to withstand the unavoidable comparison among all existing algorithms already bearing fruit in various theories, practices, and applications. As a newborn algorithm, it should undoubtedly carry extra advantages for use when other algorithms could fail or have a low capacity for compliance. More important, it induces no cost or has less cost in computation.

In the context of evolutionary computing, multiobjective optimization in particular is a challenging task for obtaining an adequate solution to furnish all the necessary design specifications. The contradiction between convergence and diversity is the main hindrance that can vary from one design to the next. Some are better performers in certain specific functions, while others might not even be up to the task. This book highlights the historical background of multiobjective optimization. This includes the most popular algorithms in use and then progressively leads to the forefront objective of the jumping gene algorithm.

As a new scientific development for which there previously was no available literature to substantiate the claim, the translation of the physical phenomenon of jumping gene transposition into a computationally pro-grammable language is not only essential but also difficult by all standards. A comprehensive review of all the jumping gene operations was therefore necessary, as was a crucial effort for realistic realization if it can be success-fully applied to future engineering designs.

After establishing the computational language for jumping gene operations, the next difficult task is to verify it mathematically. This is a vigorous exercise if the whole concept of the algorithm can be placed among the leading evolutionary algorithms. Holland's model was the first schema theorem used to illustrate the operations of genetic algorithms, but it practically fell short of predicting the behavior of the algorithm over multiple generations. It may not even be meaningful if only a lower bound condition is considered. Instead, Stephen and Waelbroeck's model, which gives an exact formulation rather than a lower-bound model, was duly applied.

The jumping gene operations, namely, copy-and-paste and cut-and-paste, were then explicitly studied based on their mathematical formulations. An equilibrium theorem was created to justify the dynamic analysis of these functional models with backup of appropriate but mindful simulation runs. To demonstrate the usefulness of the algorithm over other evolutionary algo-rithms, the performance-measured metrics in the domains of convergence and diversity were therefore thoroughly compared and examined.

With all these fundamentals firmly established and reinforced both con-ceptually and scientifically, the ultimate test was to realize the algorithm in practical use. However, it should be stressed that the practical implementa-tions presented in this book have a specific connotation in mind. That is, they showcase the ultimate superiority of the algorithm purposefully in terms of both convergence and diversity measures, as these are the undeniable crite-ria for any multiobjective algorithm to use for comprehension. These typical applications should not necessarily be limited thus far. For other engineering-related systems, for which the objectives of the design are conflicting or when the outlier solutions are desired, this newly developed jumping gene algo-rithm can aptly play an influential role in meeting such a challenging task.

K. S. Tang, T. M. Chan, R. J. Yin, and K. F. Man
Department of Electronic Engineering
City University of Hong Kong

About the Authors

Kit Sang Tang received his BSc from the University of Hong Kong in 1988 and his MSc and PhD from City University of Hong Kong in 1992 and 1996, respectively. He is currently an associate professor in the Department of Electronic Engineering at City University of Hong Kong. He has published over 90 journal papers and five book chapters and has coauthored two books, focusing on genetic algorithms and chaotic theory.

Tak Ming Chan received his BSc in applied physics from Hong Kong Baptist University in 1999 and his MPhil and PhD in electronic engineering from City University of Hong Kong in 2001 and 2006, respectively. He was a research associate in the Department of Industrial and Systems Engineering at the Hong Kong Polytechnic University from 2006 to 2007 and a postdoctoral fellow in the Department of Production and Systems Engineering, University of Minho, Portugal from 2007 to 2009.

Richard Jacob Yin obtained his BEng in information technology in 2004 and his PhD in electronic engineering in 2010 from the City University of Hong Kong. He is now an electronic engineer at ASM Assembly Automation Hong Kong Limited.

Kim Fung Man is a chair professor and head of the Electronic Engineering Department at City University of Hong Kong. He received his PhD from Cranfield Institute of Technology, UK. He is currently the co-editor-in-chief of *IEEE Transactions of Industrial Electronics*. He has coauthored three books and published extensively in his field of interest.

1

Introduction

1.1 Background on Genetic Algorithms

The genetic algorithm (GA) was first proposed by John Holland [30] in the late 1960s. As a major family in evolutionary algorithms, it was inspired by the natural phenomenon of biological evolution, relying on the principle of "survival of the fittest." The basic flowchart of a conventional GA is given in Figure 1.1. A GA process begins with a random population in which potential solutions are encoded into chromosomes. By evaluating each chromosome against the objective functions, its goodness, represented by a fitness value, can be determined. Based on the fitness value, some chromosomes are then selected to form a mating pool that will undergo genetic operations, namely, crossover and mutation, to produce some new solutions. The chromosomes (called the parents) in the mating pool exchange their genes to generate new chromosomes via crossover, while some genes are changed by a random process called mutation. These two operations keep the balance between the exploration and exploitation of solutions so that a GA can form better solutions based on acquired information and some random processes. The newly formed chromosomes are then assessed by objective functions, and fitness values are assigned. Usually, fitter offspring will replace some or all of the old ones such that a new population is obtained. This genetic cycle is repeated until some criteria are met. Further operational details of GA can be found in [41].

After about 30 years of development worldwide, the GA has been successfully applied to various real-world applications [3]. These areas include physics, chemistry, engineering, medicine, business and management, finance and trade, and more. Table 1.1 presents a summary of the work. This list is by no means exhaustive, but it should provide an impression of how GAs have influenced the scientific world through research and applications.

According to the nature of the problems, these applications can also be classified as single- or multiple-objective optimization. For single-objective optimization, only one objective is targeted, while there are several in multiobjective optimization. A unique global solution is commonly expected in the former, but a set of global solutions should result from the latter.

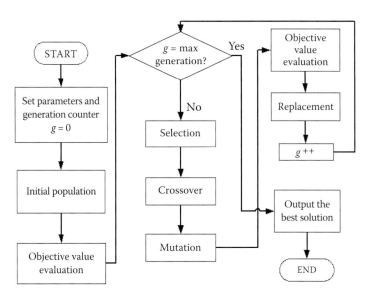

FIGURE 1.1
Genetic cycle of a GA. (From Yin, J.J., Tang, K.S., Man, K.F., A comparison of optimization algorithms for biological neural network identification, *IEEE Transactions on Industrial Electronics*, 57(3), 1127–1131, 2010.)

TABLE 1.1

Typical Applications of GA in Different Research Areas

Research Area	Examples of Applications
Physics	Water reactor [49]; two-nucleon knockout reactions [32]; Tokamak poloidal field configuration [2]; four-level triplicator [6]; spectrum assignment [51]
Chemistry	Biological activity of combinatorial compound libraries [52]; catalytic reaction sets [39]; molecule conformational search [35]; modeling of response surfaces in high-performance liquid chromatography [47]; structure elucidation of molecules [46]
Engineering	Network design [38]; city planning [4]; aerodynamic design [48]; filter design [43]; path planning for underwater robot [1]
Medicine	Allocation of radiological worker [8]; medical imaging [28]; treatment planning [9]; classification of traditional Chinese medicine [54]; medical diagnosis [45]
Business and management	Personnel scheduling [19]; forest management [23]; job shop scheduling [33]; unit commitment [14]; distribution [7]
Finance and trading	Performance prediction for individual stock [40]; financial time series [50]; economic models [42]; investment strategies [34]; trade strategies [31]

By transforming multiple objectives into a single fitness value, as a population-based optimization algorithm, the GA is especially effective for solving multiobjective optimization problems (MOPs) [25,26]. A number of multiobjective evolutionary algorithms (MOEAs) have been proposed, including the multiobjective genetic algorithm (MOGA) [25]; niched Pareto genetic algorithm 2 (NPGA2) [20]; nondominated sorting genetic algorithm 2 (NSGA2) [17,18]; strength Pareto evolutionary algorithm 2 (SPEA2) [55]; Pareto archived evolution strategy (PAES) [36,37]; microgenetic algorithm (MICROGA) [11,12]; and so on.

Except for the PAES, all of these MOEAs are GA based, and they rely on Pareto sampling techniques, which are capable of generating multiple solutions in a single simulation run. However, true Pareto-optimal solutions are seldom reached by MOEAs. A set of nondominated solutions is thus obtained instead [10,16], and this solution set is preferably as close to the true Pareto-optimal front or reference front [10] as possible. Then, selecting a compromising solution for a particular application is the responsibility of the decision maker.

Although advanced developments of various MOEAs together with many additional measures have been suggested (e.g., mating restrictions [5,21,22,29,30], fitness sharing [27], and the crowding scheme [15]), it is still not an easy task to obtain a widespread nondominated solution set, and the convergence speed is usually slow. Thus, another biological genetic phenomenon is considered as a possible way to improve search performance for evolutionary computation, which is also the focus of this book.

When evolution proceeds generation by generation in an MOEA, the genes in the chromosomes of the parent are passed to the offspring. This is called vertical gene transmission (VGT). In the early period, it was thought that all genes in chromosomes were fixed and could be transferred only through the VGT process. Nevertheless, this was not true; biologist Barbara McClintock first discovered that some kinds of genes could "jump" horizontally in the same chromosome or to other chromosomes in the same generation when there was stress exerted on the chromosomes [13,24,44]. Therefore, these genes were called jumping genes (JGs); this phenomenon was termed horizontal gene transmission (HGT).

The framework of an MOEA provides a unique platform for the imitation of the JG as each fitness function can affect a certain part (genes) of the chromosomes throughout the evolutionary process. The striking findings of the JG are that it not only offers better convergence but also presents a wider spread of nondominated solutions, especially at both ends of the true Pareto-optimal front.

This book thus serves as a key reference for the introduction of the JG in evolutionary computation, covering the fundamental concept, theoretical justification, simulation verifications, and applications of the JG.

1.2 Organization of Chapters

This book has nine chapters, including the introduction in this chapter. In Chapter 2, a thorough review of state-of-the-art multiobjective optimization techniques is presented. This leads the way for new development of evolutionary theory such as JG to further unravel the shortfall of multiobjective optimization, particularly when convergence and diversity are both in demand.

Considering the fact that the biogenetic phenomenon of the JG may be new to readers, the basic biological gene transposition process is described in Chapter 3. It serves as a shortcut to the bioscience for comprehension of the JG concept. The transformation of gene manipulation in the computational forms, namely, copy and paste and cut and paste, is then presented, supported by a series of studies on its suitability.

As the JG is such a new algorithm, it is preferable to derive the necessary proofs of the JG mathematically at the beginning. A number of theoretical proposals in this endeavor have been under consideration, but it was the schema theory deployed by Stephens and Waelbroeck's model that became the winner. Chapter 4 provides an exact mathematical formulation for the growth of the schemata; consequently, two delightful mathematical equations, one for the copy-and-paste and the other for the cut-and-paste operations are established and presented. Further efforts to derive important theories for understanding the dynamics of the growth of schemata, and methods have been established for verifying all the analytical findings. All these results are given in Chapter 4.

Knowing that solutions for an MOP can be measured by their convergence and diversity, a number of suitable metrics are explained in Chapter 5. All these were statistically conducted based on a range of benchmark mathematical functions, whether constrained or unconstrained. The JG was singled out in most of the tests, as clearly illustrated in Chapter 5.

This successful work in JG development resulted in high confidence, naturally leading to the area of practical applications. Chapter 6 addresses the first application, in which the radio-to-fiber repeater placement in wireless local loop systems was studied. In Chapter 7, another application, a resource management problem in the wideband code division multiple access (WCDMA) system, is given. The final application discussed is base station placement in wireless local-area networks, with details given in Chapter 8. The problem is tackled of placing base stations in appropriate positions to provide sufficient radio coverage quality for each terminal. It is a typical minimax optimization problem for which adequate solutions may not be easily obtained.

All these examples demonstrated that the JG can indeed be a new but useful addition to the existing evolutionary algorithms when a better result in trading off the measure between convergence and diversity performance metrics is desired. Finally, a short summary is given in Chapter 9 regarding all the work done to date on this topic.

References

1. Alvarez, A., Caiti, A., Onken, R., Evolutionary path planning for autonomous underwater vehicles in a variable ocean, *IEEE Journal of Oceanic Engineering*, 29(2), 418–429, 2004.
2. Amrollahi, R., Minoo, H., Khorasani, S., Dini, F., Optimization of tokamak poloidal field configuration by genetic algorithms, in *Proceedings of 18th IAEA Fusion Energy Conference* (CD-ROM), Sorrento, Italy, October 2000.
3. Bäck, T., Fogel, D. B., Michalewicz, Z. (Eds.), *Handbook of Evolutionary Computation*, Bristol, UK: Institute of Physics, 1997.
4. Balling, R. J., Taber, J. T., Brown, M. R., Day, K., Multiobjective urban planning using a genetic algorithm, *ASCE Journal of Urban Planning and Development*, 152(2), 86–99, 1999.
5. Booker, L.B., Improving the performance of genetic algorithms in classifier systems, in *Proceedings of the First International Conference on Genetic Algorithms and Their Applications*, Pittsburgh, July 1985.
6. Borghi, R., Frezza, F., Pajewski, L., Santarsiero, M., Schettini, G., Optimization of a four-level triplicator using genetic algorithms, *Journal of Electromagnetic Waves and Applications*, 15(9), 1161–1174, 2001.
7. Bredström, D., Carlsson, D., Rönnqvist, M., A hybrid algorithm for distribution problems, *IEEE Intelligent Systems*, 20(4), 19–25, 2005.
8. Chen, Y., Narita, M., Tsuji, M., Sa, S., A genetic algorithm approach to optimization for the radiological worker allocation problem, *Health Physics*, 70(2), 180–186, 1996.
9. Chong, J. L., Ahmad, S. U., Leyman, A. R., Meng, H. E., A multi-modal approach to radiotherapy treatment planning optimization, in *Proceedings of 22nd Annual International Conference of the IEEE Engineering in Medicine and Biology Society*, Chicago, 2000, 3:1834–1836.
10. Coello Coello, C. A., Van Veldhuizen, D. A., Lamont, G. B., *Evolutionary Algorithms for Solving Multi-objective Problems*, New York: Kluwer Academic, 2002.
11. Coello Coello, C. A., Toscano Pulido, G., A micro-genetic algorithm for multiobjective optimization, in *Proceedings of the First International Conference on Evolutionary Multi-Criterion Optimization*, Zurich, March 2001, 126–140.
12. Coello Coello, C. A., Toscano Pulido, G., Multiobjective optimization using a micro-genetic algorithm, in *Proceedings of the Genetic and Evolutionary Computation Conference*, San Francisco, August 2001, 274–282.
13. Cohen, S. N., Shapiro, J. A., Transposable genetic elements, *Scientific American*, 242(2), 36–45, 1980.
14. Damousis, G., Bakirtzis, A. G., Dokopoulos, P. S., A solution to the unit-commitment problem using integer-coded genetic algorithm, *IEEE Transactions on Power Systems*, 19(2), 1165–1172, 2004.
15. De Jong, K. A., An Analysis of the Behavior of a Class of Genetic Adaptive Systems, Ph.D. thesis, University of Michigan, 1975.
16. Deb, K. *Multi-objective Optimization Using Evolutionary Algorithms*, London: Wiley, 2001.
17. Deb, K., Agrawal, S., Pratap, A., Meyarivan, T., A fast elitist non-dominated sorting genetic algorithm for multi-objective optimization: NSGA-II, in *Proceedings*

of the Sixth International Conference on Parallel Problem Solving from Nature, Paris, September 2000, 849–858.

18. Deb, K., Pratap, A., Agrawal, S., Meyarivan, T., A fast and elitist multiobjective genetic algorithm: NSGA-II, *IEEE Transactions on Evolutionary Computation*, 6(2), 182–197, 2002.

19. Easton, F., Mansour, N., A distributed genetic algorithm for deterministic and stochastic labor scheduling problems, *European Journal of Operational Research*, 118(3), 505–523, 1999.

20. Erickson, M., Mayer, A., Horn, J., The niched Pareto genetic algorithm 2 applied to the design of groundwater remediation systems, in *Proceedings of the First International Conference on Evolutionary Multi-Criterion Optimization*, Zurich, March 2001, 681–695.

21. Eshelman, L. J., The CHC adaptive search algorithm: How to have safe search when engaging in nontraditional genetic recombination, in *Proceedings of First Workshop on Foundations of Genetic Algorithms*, Bloomington Campus, IN, July 1990, 265–283.

22. Eshelman, L. J., Schaffer, J. D., Preventing premature convergence in genetic algorithms by preventing incest, in *Proceedings of Fourth International Conference on Genetic Algorithms*, San Diego, July 1991, 115–122.

23. Falcão, O., Borges, J. G., Designing an evolution program for solving integer forest management scheduling models: An application in Portugal, *Forest Science*, 47(2), 158–168, 2001.

24. Fedoroff, N., Botstein, D. (Eds.), *The Dynamic Genome: Barbara McClintock's Ideas in the Century of Genetics*, New York: Cold Spring Harbor Laboratory Press, 1992.

25. Fonseca, C. M., Fleming, P. J., Multiobjective optimization and multiple constraint handling with evolutionary algorithms—Part I: A unified formulation, *IEEE Transactions on System, Man and Cybernetic Part A: Systems and Humans*, 28(1), 26–37, 1998.

26. Fonseca, C. M., Fleming, P. J., Genetic algorithms for multiobjective optimization: formulation, discussion and generalization, in *Proceedings of the Fifth International Conference on Genetic Algorithms*, Fairfax, Virginia, June 1989, 2–9.

27. Goldberg, D. E., Richardson, J., Genetic algorithms with sharing for multimodal function optimization, *Proceedings of the Second International Conference on Genetic Algorithms*, Cambridge, MA, July 1987, 41–49.

28. Gudmundsson, M., El-Kwae, E. A., Kabuka, M. R., Edge detection in medical images using a genetic algorithm, *IEEE Transactions on Medical Imaging*, 17(3), 469–474, 1998.

29. Hillis, W. D., Co-evolving parasites improve simulated evolution as an optimization procedure, in *Proceedings of the Ninth Annual International Conference of the Center for Nonlinear Studies on Self-organizing, Collective, and Cooperative Phenomena in Natural and Artificial Computing Networks on Emergent Computation*, Los Alamos, NM, May 1989, 228–234.

30. Holland, J. H., *Adaptation in Natural and Artificial Systems*, Ann Arbor, MI: MIT Press, 1975.

31. Hryshko, A., Downs, T., An implementation of genetic algorithms as a basis for a trading system on the foreign exchange market, in *Proceedings of Congress on Evolutionary Computation*, Canberra, December 2003, 3, 1695–1701.

32. Ireland, G., Using a genetic algorithm to investigate two-nucleon knockout reactions, *Journal of Physics G: Nuclear and Particle Physics*, 26(2), 157–166, 2000.

33. Jensen, M. T., Generating robust and flexible job shop schedules using genetic algorithms, *IEEE Transactions on Evolutionary Computation*, 7(3), 275–288, 2003.
34. Jiang, R., Szeto, K. Y., Extraction of investment strategies based on moving averages: A genetic algorithm approach, in *Proceedings of IEEE International Conference on Computational Intelligence for Financial Engineering*, Hong Kong, March 2003, 403–410.
35. Kariuki, B. M., Calcagno, P., Harris, K. D. M., Philp, D., Johnston, R. L., Evolving opportunities in structure solution from powder diffraction data—crystal structure determination of a molecular system with twelve variable torsion angles, *Angewandte Chemie*, 38(6), 831–835, 1999.
36. Knowles, J. D., Corne, D. W., The Pareto archived evolution strategy: A new baseline algorithm for Pareto multiobjective optimization, in *Proceedings of Congress on Evolutionary Computation*, Washington, DC, 1999, 1:98–105.
37. Knowles, J. D., Corne, D. W., Approximating the nondominated front using the Pareto archived evolution strategy, *Evolutionary Computation*, 8(2), 149–172, 2000.
38. Ko, K. T., Tang, K. S., Chan, C. Y., Man, K. F., Kwong, S., Using genetic algorithms to design mesh networks, *Computer*, 30(8), 56–61, 1997.
39. Lohn, J. D., Colombano, S. P., Scargle, J., Stassinopoulos, D., Haith, G. L., Evolving catalytic reaction sets using genetic algorithms, in *Proceedings of IEEE International Conference on Evolutionary Computation*, Anchorage, May 1998, 487–492.
40. Mahfoud, S., Mani, G., Financial forecasting using genetic algorithms, *Applied Artificial Intelligence*, 10(6), 543–565, 1996.
41. Man, K. F., Tang, K. S., Kwong, S., Genetic algorithms: concepts and applications, *IEEE Transactions on Industrial Electronics*, 43(5), 519–534, 1996.
42. Mardle, S. J., Pascoe, S., Tamiz, M., An investigation of genetic algorithms for the optimization of multi-objective fisheries bioeconomic models, *International Transactions in Operational Research*, 7(1), 33–49, 2000.
43. Mastorakis, N. E., Gonos, I. F., Swamy, M. N. S., Design of two-dimensional recursive filters using genetic algorithms, *IEEE Transactions on Circuits and Systems I: Fundamental Theory and Applications*, 50(5), 634–639, 2003.
44. McClintock, B., Chromosome organization and genic expression, *Cold Spring Harbor Symposia on Quantitative Biology*, 16, 13–47, 1951.
45. Meesad, P., Yen, G. G., Combined numerical and linguistic knowledge representation and its application to medical diagnosis, *IEEE Transactions on Systems, Man and Cybernetics—Part A: Systems and Humans*, 33(2), 206–222, 2003.
46. Meiler, J., Will, M., Genius: A genetic algorithm for automated structure elucidation from 13C NMR spectra, *Journal of the American Chemical Society*, 124(9), 1868–1870, 2002.
47. Nikitas, P., Pappa-Louisi, A., Papageorgiou, A., Zitrou, A., On the use of genetic algorithms for response surface modeling in high-performance liquid chromatography and their combination with the Microsoft solve, *Journal of Chromatography A*, 942, 93–105, 2002.
48. Obayashi, S., Tsukahara, T., Nakamura, T., Multiobjective genetic algorithm applied to aerodynamic design of cascade airfoils, *IEEE Transactions on Industrial Electronics*, 47(1), 211–216, 2000.
49. Parks, T., Multiobjective pressurized water reactor reload core design by nondominated genetic algorithm search, *Nuclear Science and Engineering*, 124(1), 178–187, 1996.

50. Ruspini, H., Zwir, I. S., Automated qualitative description of measurements, in *Proceedings of 16th IEEE Instrumentation and Measurement Technology Conference*, Venice, May 1999, 2:1086–1091.
51. Schmitt, M., Ratzer, C., Kleinermanns, K., Meerts, W. L., Determination of the structure of 7-azaindole in the electronic ground and excited state using high-resolution ultraviolet spectroscopy and an automated assignment based on a genetic algorithm, *Molecular Physics*, 102(14–15), 1605–1614, 2004.
52. Weber, L., Wallbaum, S., Broger, C., Gubernator, K., Optimization of the biological activity of combinatorial compound libraries by a genetic algorithm, *Angewandte Chemie*, 34, 2280–2282, 1995.
53. Yin, J. J., Tang, K. S., Man, K. F., A comparison of optimization algorithms for biological neural network identification, *IEEE Transactions on Industrial Electronics*, 57(3), 1127–1131, 2010.
54. Zhang, L., Zhao, Y., Yang, Z., Wang, J., Cai, S., Liu, H., Classification of traditional Chinese medicine by nearest-neighbour classifier and genetic algorithm, in *Proceedings of Fifth International Conference on Information Fusion*, Annapolis, MD, July 2002, 2:1596–1601.
55. Zitzler, E., Laumanns, M., Thiele, L., *SPEA2: Improving the Strength Pareto Evolutionary Algorithm*, Technical Report (TIK-Report 103), Zurich: Swiss Federal Institute of Technology (ETH), 2001.

2

Overview of Multiobjective Optimization

2.1 Classification of Optimization Methods

Optimization methods for solving multiobjective optimization problems (MOPs) can be classified into three categories: enumerative methods, deterministic methods, and stochastic methods [12,51,58].

2.1.1 Enumerative Methods

The enumerative method is a simple and straightforward searching scheme. It inspects every solution in the search space. As a consequence, it is guaranteed to obtain true Pareto-optimal solutions. However, when the search space (i.e., the total number of solutions) is huge, it is computationally infeasible to list all the solutions. Accordingly, this method is only practical for problems with a small search space. Unfortunately, the search space of many problems is large, especially when real-world applications are tackled.

2.1.2 Deterministic Methods

To deal with problems having enormous search spaces, deterministic methods have been suggested. These methods incorporate the problem domain knowledge (heuristics) to limit the search space so that acceptable solutions can be found within a reasonable time [51]. Some typical examples of deterministic methods are given in Table 2.1.

Deterministic methods have been successfully adopted for solving a wide variety of problems [10,54]. Nevertheless, they face difficulties when coping with high-dimensional, multimodal, and nondeterministic polynomial-time (NP) complete MOPs. It is because their search performances largely rely on the implemented method, which in turn is dependent on the nature of the error surface of the search space [29,33,39,50,51].

Another disadvantage occurs because the deterministic method is basically a point-by-point approach [19]. In a single simulation run, only one

TABLE 2.1

Examples of Deterministic
Optimization Methods

Deterministic Methods
Greedy [9, 18, 45]
Hill climbing [61]
Branch and bound [11,27,43]
Depth first [56]
Breadth first [56]
Best first [56]
Calculus based [2]
Mathematical programming [41,53]

optimized solution is located. Therefore, if a set of optimal or suboptimal solutions is desired, such as when handling MOPs, multiple runs are needed.

2.1.3 Stochastic Methods

Rather than using enumerative and deterministic methods for solving MOPs, stochastic methods are more useful [12]. Stochastic methods can acquire multiple solutions in a single simulation run and have their own schemes to prevent trapping in the local optima [19].

Usually, an encode/decode mapping mechanism between the problem and algorithm domains is needed. Furthermore, a specific function is required to reflect the goodness of the solutions. Production of true Pareto-optimal solutions is not guaranteed. Nonetheless, they can offer reasonably good solutions to MOPs for which deterministic methods fail [39,44]. Some typical examples of stochastic methods are listed in Table 2.2.

TABLE 2.2

Examples of Stochastic Optimization Methods

Stochastic Methods
Random search/walk [64,68]
Monte Carlo [34,59]
Simulated annealing [49,60]
Tabu search [35–38]
Swarm algorithms [15–17,46]
Evolutionary computation/algorithms [5,6,24,28]

2.2 Multiobjective Algorithms

The following summarizes some major stochastic optimization methods widely used for MOPs, including evolutionary algorithms, swarm algorithms, and the tabu search (TS).

Evolutionary algorithms emulate the process of natural selection for which the philosopher Herbert Spencer coined the phrase "survival of the fittest" [7,52,67]. There are in general three proposed classes: evolutionary programming, evolution strategies, and genetic algorithms (GAs). They share a number of common properties and genetic operations; their main differences are listed in Table 2.3. Our focus is on the algorithms addressing MOPs; hence, more details are given for the distinct features found in major multiobjective evolutionary algorithms (MOEAs). Most of these MOEAs are GA based; however, similar operations can also be applied for evolutionary programming and evolution strategies. Details of evolutionary programming and evolution strategies can be obtained in [3–5,8,25,28–30,61,62].

In addition to evolutionary algorithms, two main classes of swarm algorithms, namely, the ant colony [16,17] and the particle swarm optimizations (PSOs) [46], are commonly used for solving optimization problems. They are also inspired by the natural world, but their searching performances rely on a decentralized group of agents, for which their local interactions lead to a certain level of global search. Last, the TS, which was proposed by Fred Glove, is also briefly introduced. It acts as an effective algorithm for many problems for which solutions are searched by repeatedly moving from the current one to the best of its neighbors.

2.2.1 Multiobjective Genetic Algorithm

The multiobjective genetic algorithm (MOGA) was proposed by Fonseca and Fleming [31,32]. It has three features: a modified ranking scheme, modified fitness assignment, and niche count. The flowchart of the MOGA is shown in Figure 2.1.

A modified ranking scheme [32], which is slightly different from the one proposed by Goldberg, is used in the MOGA. Figure 2.2 depicts such a

TABLE 2.3

Differences between Three Evolutionary Algorithms

Evolutionary Algorithm	Representation	Genetic Operators
Evolutionary programming	Real values	Mutation and $(\mu + \lambda)$ selection
Evolution strategies	Real values and strategy parameters	Crossover, mutation, and $(\mu + \lambda)$ or (μ, λ) selection where μ and λ are the number of parents and children, respectively
Genetic algorithms	Binary or real values	Crossover, mutation, and selection

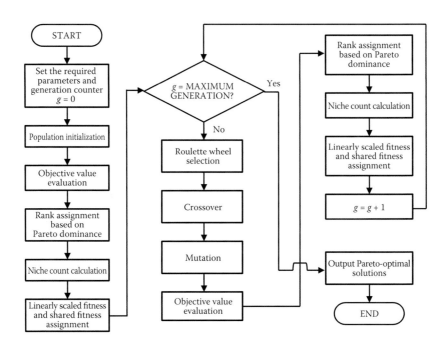

FIGURE 2.1
Flowchart of MOGA.

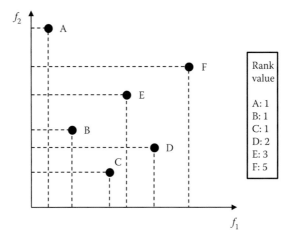

FIGURE 2.2
Fonseca and Fleming's ranking scheme for minimization problems.

scheme. The rank of an individual i is given by the number of individuals q that dominate i plus one, that is,

$$\text{Rank}(i) = 1 + q \tag{2.1}$$

where the definition of domination can be mathematically described as follows:

For minimization problems, chromosome i is dominated by chromosome j (denoted as $i \succ j$) if and only if

$$f_m(i) \geq f_m(j) \; \forall m \quad \text{and} \quad f_{m'}(i) > f_{m'}(j) \; \exists m' \tag{2.2}$$

where f_m is the mth objective function. A similar definition can be derived for maximization problems.

2.2.1.1 Modified Fitness Assignment

The fitness of an individual is assigned as follows:

1. Sort the population according to the ranks.
2. Assign fitness to individuals by interpolating from the best (rank 1) individual to the worst (rank $n \leq N$, where N is the population size) on the basis of some functions, usually linear or exponential, but other types are possible.
3. Average the fitness assigned to individuals with the same rank so that all of them are sampled with the same probability.

2.2.1.2 Fitness Sharing

In general, it is difficult to obtain a uniformly distributed and widespread set of nondominated solutions in multiobjective optimization. This is caused by the stochastic selection errors in a finite population size and the convergence of the population to a small region of Pareto-optimal front (i.e., a phenomenon called genetic drift occurs in both natural and artificial evolution). Fitness sharing was suggested [31] to try to resolve these problems. It utilizes the concept of individual competition for finite resources in a closed environment. Individuals reduce each other's fitness according to their similarity, which can be measured in different domain spaces (e.g., the Hamming distance of the genotype of the chromosome, distance between the objective values, etc.).

To perform fitness sharing, a sharing function is required to determine the fitness reduction of a chromosome based on the crowding degree caused by its neighbors. The commonly used sharing function is given by

$$sf(d_{ij}) = \begin{cases} 1 - \left(\dfrac{d_{ij}}{\sigma_{share}} \right)^{\alpha} & \text{if } d_{ij} < \sigma_{share} \\ 0 & \text{otherwise} \end{cases} \tag{2.3}$$

where α is a constant to determine the shape of the sharing function, σ_{share} is the niche radius chosen by the user for the minimal separation desired, and d_{ij} is the distance specifying the similarity of two individuals.

The shared fitness of a chromosome is computed by

$$f'(i) = \frac{f(i)}{c_i} \tag{2.4}$$

and c_i is the niche count, defined as

$$c_i = \sum_{j=1}^{N} sf(d_{ij}) \tag{2.5}$$

where N is the population size.

2.2.2 Niched Pareto Genetic Algorithm 2

The niched Pareto genetic algorithm 2 (NPGA2), which is an improved version of the NPGA [42], was suggested by Erickson et al. [26]. The flowchart of NPGA2 is shown in Figure 2.3. The special feature of this algorithm is the use of fitness sharing when the tournament selection ends in a tie. A tie means that both picked individuals are dominated or nondominated. If a tie

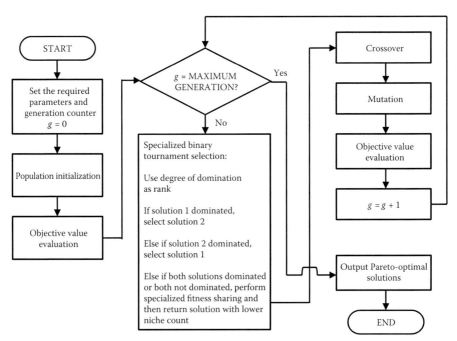

FIGURE 2.3
Flowchart of NPGA2.

happens, their niche counts c_i in Equation (2.5) are computed, and the one with a lower niche count will be the winner and included in the mating pool.

Oei et al. [55] showed that the combination of tournament selection and fitness sharing would lead to chaotic perturbations of the population composition. Therefore, a method called continuously updated fitness sharing was suggested. Niche counts are calculated by using individuals in the partially filled next-generation population rather than the current-generation population. To facilitate the calculation of niche count, the values of different objective functions should be normalized, that is,

$$f'_m = \frac{f_m - f_m^{(\min)}}{f_m^{(\max)} - f_m^{(\min)}} \tag{2.6}$$

where f'_m is the normalized objective value of f_m, $f_m^{(\min)}$ and $f_m^{(\max)}$ are the minimum and maximum values of the mth objective function, respectively.

2.2.3 Nondominated Sorting Genetic Algorithm 2

The nondominated sorting genetic algorithm 2 (NSGA2) is an enhanced version of the NSGA [65] proposed by Deb et al. [20,21], and its flowchart is given in Figure 2.4. It has four peculiarities: fast nondominated sorting, crowding distance assignment, crowded comparison operator, and elitism strategy.

2.2.3.1 Fast Nondominated Sorting Approach

The design of the fast nondominated sorting approach in NSGA2 is aimed at reducing the high computational complexity of the traditional nondominated sorting approach, which is $O(MN^3)$.

In the traditional nondominated sorting approach, each solution has to be compared with every other solution in the population to determine which lies on the first nondominated front. For a population of size N, $O(MN)$ comparisons are needed for each solution, and the total complexity is $O(MN^2)$ for searching all solutions of the first nondominated front where M is the total number of objectives.

To obtain the second nondominated front, the solutions of the first front are temporarily discounted and the procedure is performed. In the worst case, locating the second front also requires $O(MN^2)$ computations. A similar calculation applies to the other (i.e., third, fourth, and so on) nondominated fronts. Hence, in the worst case, overall $O(MN^3)$ computations are needed.

On the other hand, the fast nondominated sorting approach only requires an overall computational complexity of $O(MN^2)$. Referring to the flowchart shown in Figure 2.5, its operations are given as follows: For each solution

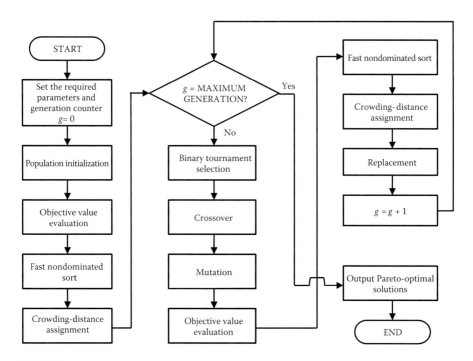

FIGURE 2.4
Flowchart of NSGA2.

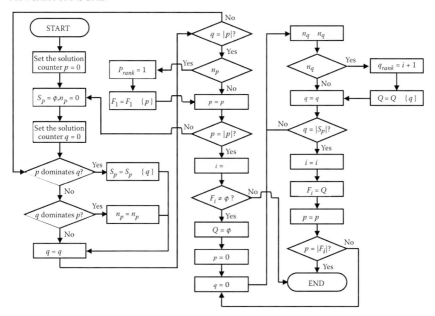

FIGURE 2.5
Flowchart of the fast nondominated sorting approach.

$p \in P$, where P is the population with size $|P|$, the number of solutions that dominate p (called the domination count n_p) and a set of solutions that p dominates (denoted S_p) are computed. This requires $O(MN^2)$ comparisons. Then, solutions with their $n_p = 0$ are the members of the first nondominated front.

For every solution p lying on the first front, its values of n_q and S_p are decreased by one. If there exists any solution q with $n_q = 0$, it is stored in a separate list Q, which contributes to the second nondominated front. This procedure is repeated until all nondominated fronts are acquired.

2.2.3.2 Crowded-Comparison Approach

Recognizing the difficulties of specifying the sharing parameter for the preservation of population diversity, a crowded-comparison approach is proposed in NSGA2. It involves two processes: the crowding-distance assignment and the crowded-comparison operation.

2.2.3.2.1 Crowding-Distance Assignment

To obtain diversified nondominated solutions, the crowding-distance assignment is devised. Its purpose is to estimate the density of solutions surrounding a particular solution within the population. Solutions in the less-crowded area will be chosen on the basis of their assigned crowding distances.

The crowding distance measures the perimeter of the cuboid formed using the nearest neighbors as the vertices. An example is depicted in Figure 2.6. The crowding distance of the ith nondominated solution marked with the solid circle is the average side length of the cuboid represented by a dashed

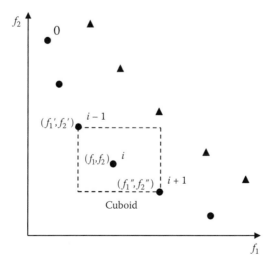

FIGURE 2.6
Crowding-distance computation.

box. Assuming that the nondominated solutions i, $(i-1)$, and $(i+1)$ in Figure 2.6 possess the objective values of (f_1, f_2), (f_1', f_2'), and (f_1'', f_2''), respectively, the crowding distance $I[i]_{distance}$ of the solution i in the nondominated set I is given by

$$I[i]_{distance} = |f_1' - f_1| + |f_1'' - f_1| + |f_2'' - f_2'| \qquad (2.7)$$

The flowchart of the crowding-distance assignment is shown in Figure 2.7. Its procedures are as follows:

1. Initialize the crowding distance $I[i]_{distance}$ of each nondominated solution to zero.
2. Set the counter of objective function to one, $m = 1$.
3. Sort the population according to each objective value of the mth objective function in ascending order.
4. For the mth objective function, assign a nominated relatively large value to $I[i]_{distance}$ of the extreme solutions of the nondominated front, that is, the smallest and largest function values. This value must be greater than the crowding distance values of other solutions within the same nondominated front.

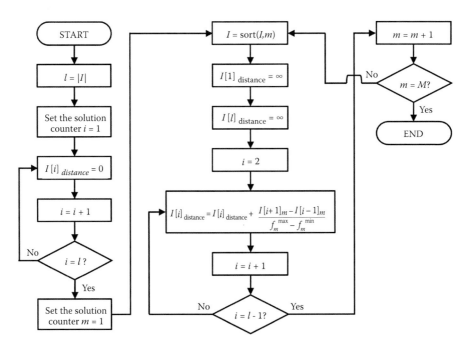

FIGURE 2.7
Flowchart of the crowding-distance assignment.

5. Calculate $I[i]_{distance}$ for each of the remaining nondominated solutions $i = 2,3,\cdots,(l-1)$ by

$$I[i]_{distance} = I[i]_{distance} + \frac{I[i+1]_m - I[i-1]_m}{f_m^{(max)} - f_m^{(min)}} \qquad (2.8)$$

where l is the total number of solutions in the nondominated set I, $I[i]_m$ is the mth objective function value of the ith solution in I, and $f_m^{(min)}$ and $f_m^{(max)}$ are the minimum and maximum values of the mth objective function, respectively.

6. Update m by setting $m = m + 1$. If $m \leq M$ where M is the total number of objective functions, go to step 3.

A solution with a smaller crowding distance means that it is more crowded. In other words, this solution and its other surrounding solutions are close in distance.

2.2.3.2.2 Crowded-Comparison Operator

The crowded-comparison operator, implemented in the selection process, aims to obtain more uniformly distributed nondominated solutions. If any two solutions picked have different ranks, the solution with the lower rank will be chosen. However, if their ranks are the same, the solution with the larger crowding distance (i.e., located in a less-crowded region) is preferred.

2.2.3.2 Elitism Strategy

The procedure for executing elitism is outlined as follows:

1. Assuming that the tth generation is considered, combine the original population P_t and the offspring population Q_t to form a population R_t.
2. Sort R_t by the fast nondominated sorting approach and then identify different nondominated fronts $F_1, F_2, \ldots$.
3. Include these fronts in the new population P_{t+1} one by one until P_{t+1} with size N is full.

If there is insufficient room to include all members of a particular front, say F_i into P_{t+1}, the first j members of F_i with larger crowding distances will be selected to fill P_{t+1}. The flowchart of the elitism strategy is given in Figure 2.8.

2.2.4 Strength Pareto Evolutionary Algorithm 2

The strength Pareto evolutionary algorithm 2 (SPEA2) [69] is an improved version of SPEA [70] that includes three distinct features: strength value and

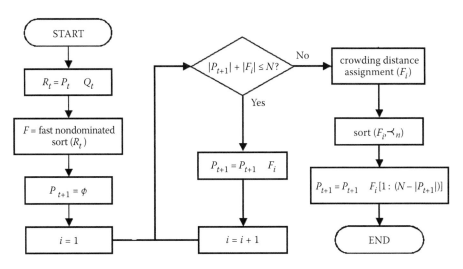

FIGURE 2.8
Flowchart of the elitism strategy.

raw fitness, density estimation, and an archive truncation method. Its flowchart is depicted in Figure 2.9.

2.2.4.1 Strength Value and Raw Fitness

Consider a population P and an archive A containing the other solutions dominated by solution i, the strength value of solution i, $S(i)$, is specified by

$$S(i) = |\{j : (j \in P + A) \wedge (i \succ j)\}| \tag{2.9}$$

$|\bullet|$ is the cardinality of a set, $+$ is the multiset union, and $\succ$ is the Pareto dominance relation defined in Equation (2.2).

The raw fitness of the individual i, $R(i)$, is acquired by summing the strength values of its dominators in both population and archive. In mathematics, one has

$$R(i) = \sum_{j \in P + A; j \succ i} S(j) \tag{2.10}$$

Assuming that a minimization problem is considered, an individual with a zero raw fitness value is a nondominated one. In contrast, an individual with a high raw fitness value means that it is dominated by many other individuals.

2.2.4.2 Density Estimation

Even though the raw fitness assignment offers a sort of niching mechanism on the basis of Pareto dominance, it may fail when most individuals do not

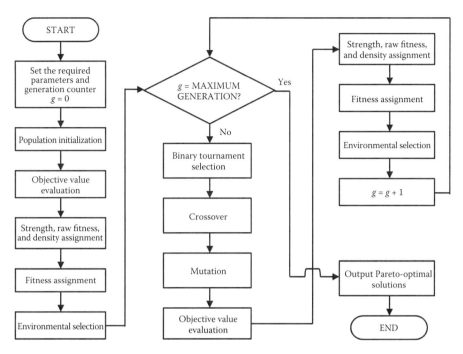

FIGURE 2.9
Flowchart of SPEA2.

dominate each other. As a result, density estimation is employed to discriminate between individuals having identical raw fitness values.

The density estimation technique used in the SPEA2 is called the kth nearest-neighbor method [63]; the density at any point is a decreasing function of the distance to the kth nearest data point, which can be expressed as

$$D(i) = \frac{1}{d_i^k + 2} \tag{2.11}$$

where $D(i)$ and d_i^k are the density and the found distance value of solution i, respectively.

A value of two is added in the denominator in Equation (2.11) to avoid division by zero and make the density value always less than one. The value of d_i^k is obtained by the following procedures:

1. The distances in objective space of all individuals j in A and P are calculated and stored in a list.

2. Sort the list in ascending order.

3. The kth element gives the value of d_i^k, where $k = \sqrt{|P| + |A|}$.

2.2.4.3 Archive Truncation Method

In each generation, the archive of the next generation is formed by copying all nondominated individuals from the archive and the population. If the number of all those nondominated individuals exceeds the archive size $|A|$, the archive truncation method is used to iteratively remove some nondominated individuals until the number reaches $|A|$.

An individual i is selected for removal if $i \leq_d j$ for all $j \in A_{t+1}$, with $i \leq_d j$ defined as

$$\forall 0 < k < |A_{t+1}| : \sigma_i^k = \sigma_j^k \ \text{ or } \ \exists 0 < k < |A_{t+1}| : \left[\left(\forall 0 < l < k \ \text{ s.t. } \ \sigma_i^l = \sigma_j^l \right) \ \text{ and } \ \sigma_i^k < \sigma_j^k \right]$$

where σ_i^k is the distance of the individual i to its kth nearest neighbor in A_{t+1}, and t is the counter of generations.

It means that an individual having the minimum distance to another individual is chosen at each stage. If there are several individuals with the minimum distance, the tie is broken by considering the second-smallest distances and so forth.

This archive truncation method can avoid the removal of boundary nondominated solutions, which is more favorable for MOPs.

2.2.5 Pareto Archived Evolution Strategy

The Pareto archived evolution strategy (PAES) is a local search-based algorithm [47,48], imitating the evolution strategy. Its flowchart is shown in Figure 2.10. In the PAES, the only genetic operation, mutation, provides a hill-climbing-like strategy. An archive with limited size is used to store the previously found nondominated solutions.

The PAES has three versions: (1 + 1)-PAES, (1 + λ)-PAES, and (μ + λ)-PAES. The first one means that a single parent generates one offspring. The second one represents a single parent producing λ offspring. The last one means that a population of μ parents generates λ offspring. In comparing these three versions, (1 + 1)-PAES has the lowest computational overhead (i.e., it is a faster algorithm) and is the simplest.

The unique characteristic of the PAES is the adaptive grid scheme. Its notion is the use of a new crowding procedure based on recursively dividing up the M-dimensional objective space to trace the crowding degrees of different regions in which the nondominated solutions in the archive fall. Its purpose is to diversify those nondominated solutions and help to remove excessive nondominated solutions located in the crowded grids (i.e., the grids with a high degree of crowding) if the number of those solutions exceeds the archive size.

The procedures of the adaptive grid scheme are outlined as follows: When each solution is generated, it is necessary to determine its grid location in the objective space. Suppose that the range of the space is defined in each

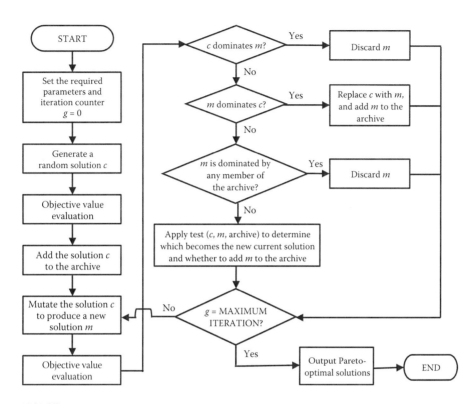

FIGURE 2.10
Flowchart of PAES.

objective; the required grid location can be acquired by repetitively bisecting the range in each objective and finding in which half the solution lies. The location of the solution is marked with a binary string of length $2^{b \times M}$ where b is the number of bisections of the space for each objective, and M is the number of objectives.

If the solution is located at the larger half of the bisection of the space, the corresponding bit in the binary string is set. To record the number and which nondominated solutions reside in each grid, a map of the grid is maintained throughout the run. Grid locations are recalculated when the range of the objective space of the archived solutions changes by a threshold amount to avoid recalculating the ranges too frequently. Only one parameter (i.e., the number of divisions of the space required) needs to be set.

2.2.6 Microgenetic Algorithm

The microgenetic algorithm (MICROGA) was suggested by Coello Coello and Toscano Pulido [13,14], and its flowchart is shown in Figure 2.11. It has three peculiarities: population memory, an adaptive grid algorithm, and three types of elitism. These are briefly described next.

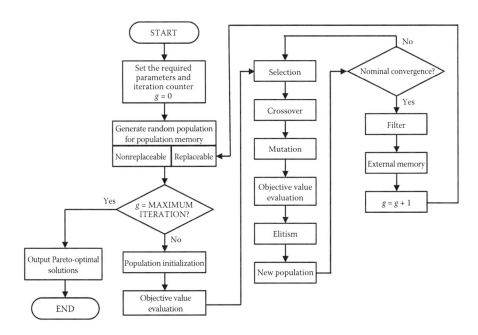

FIGURE 2.11
Flowchart of MICROGA.

2.2.6.1 Population Memory

The population memory consists of two parts, replaceable and nonreplaceable. The replaceable part may be modified after each cycle. In contrast, the non-replaceable part is unaltered during the run, aiming at providing the required diversity for the algorithm.

At the beginning of each cycle, the population is taken from both portions of the population memory so there is a mixture of randomly generated individuals (nonreplaceable) and evolved individuals (replaceable).

2.2.6.2 Adaptive Grid Algorithm

The adaptive grid algorithm employed in MICROGA is similar to that in the PAES. It offers diversity to nondominated solutions. Once the archive storing the nondominated solutions reaches its limit, the objective space covered by the archive is divided into a number of grids. Each solution in the archive is then assigned a set of coordinates.

When a new nondominated solution is generated, it is accepted if it is located at a grid where the number of stored nondominated individuals is smaller than that of the most crowded grid or located outside the previously specified boundaries. Note that this adaptive grid algorithm requires two parameters: the expected size of the Pareto front and the number of positions in which the solution space will be divided for each objective.

2.2.6.3 Three Types of Elitism

Three types of elitism scheme are implemented in MICROGA. In the first type, nondominated solutions found within the internal cycle are stored so valuable information obtained during the evolutionary process will not be lost.

For the second type, nominal solutions (i.e., the best solutions found when the nominal convergence is reached) are added to the replaceable part of the population memory to enhance the convergence of solutions. It results in a higher probability of reaching the true Pareto front over time using the crossover and mutation operations.

The last elitism scheme uniformly picks a certain number of solutions from all the regions of the Pareto front generated so far and includes them in the replaceable part of the population memory. Its purpose is to utilize the best-available solutions as the starting point, and further improvements are expected via operations (either by getting closer to the true Pareto front or by getting a better distribution).

2.2.7 Ant Colony Optimization

Ant colony optimization (ACO) is a metaheuristic optimization method that uses a model-based searching framework [16,17,23]. It has been applied for solving many hard combinatorial optimization problems [22,66].

The principle of ACO can be summarized as "a set of artificial ants moving through states of the problem corresponding to partial solutions of the problem to solve" [22], and its flowchart is shown in Figure 2.12. Its design concept follows the foraging behavior of ants. When a population of ants searches for food, the ants are sent out to explore the areas surrounding their nest in a random manner. When an ant reaches the food, it carries the food back and deposits a trail of chemical pheromone that serves as information to guide the other ants to the food source. The quantity of pheromone deposited depends on the quantity and quality of the food. After several rounds of searching, the density of the pheromone trails on the paths will indicate the shortest way to reach the food.

In computation, the agents (i.e., the ants) move with a stochastic local decision policy. The agent k moves from state i to j with a probability p_{ij}^k, which depends on two parameters:

1. The attractiveness η_{ij} of the move, indicating its a priori desirability
2. The trail level τ_{ij} of the move, indicating how proficient it has been in the past to make the particular move, hence representing a posteriori indication of the desirability of that move

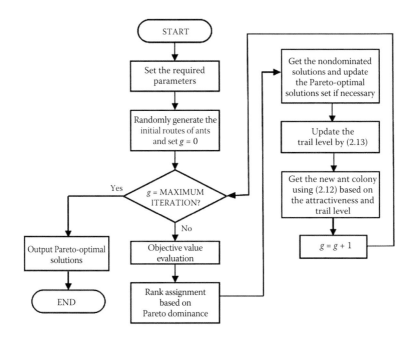

FIGURE 2.12
Flowchart of ACO.

and p_{ij}^k is given by

$$p_{ij}^k = \begin{cases} \dfrac{\tau_{ij}^\alpha + \eta_{ij}^\beta}{\displaystyle\sum_{s \notin tabu_k} (\tau_{is}^\alpha + \eta_{is}^\beta)} & \text{if}(ij) \notin tabu_k \\[3mm] 0 & \text{otherwise} \end{cases} \tag{2.12}$$

where $tabu_k$ is the unreachable move of agent k from state i; α and β represent the impacts of η_{ij} and τ_{ij}, respectively, which are user-defined; and $0 < \alpha, \beta < 1$.

When all the ants finish their moves, the trails' levels are updated by

$$\tau_{ij}(g) = \rho\tau_{ij}(g-1) + \Delta\tau_{ij} \tag{2.13}$$

where $0 < \rho < 1$ is the evaporation coefficient determined by the programmer, $\Delta\tau_{ij}$ is the sum of the contributions of all ants that have been moved from state i to j, which is computed by

$$\Delta\tau_{ij} = \sum_{k=1}^{N} \Delta\tau_{ij}^k \tag{2.14}$$

where N is the total number of ants, and $\Delta \tau_{ij}^k$ is the amount of pheromone left on the path from i to j, reflecting the quality of the solution of ant k. It can be defined as

$$\Delta \tau_{ij}^k = \begin{cases} \dfrac{Obj_k}{Q} & edge(ij) \in tour_k \\ 0 & \text{otherwise} \end{cases} \tag{2.15}$$

with a constant Q.

2.2.8 Particle Swarm Optimization

Particle swarm optimization is a swarm intelligence algorithm other than ACO. It was proposed by Ederhart and Kennedy in 1995 [46], inspired by the social action of searching for food by a flock of birds. The PSO is based on a simple but effective mechanism in which each bird, called a particle, adjusts its search direction according to three factors: its own velocity v_i, its own best previous experience ($pBest_i$), and the best experience in the flock ($gBest$) [46].

The PSO has been successful in a wide variety of optimization tasks. However, it was once considered unsuitable to deal with multiobjective optimizations until the extension made by Coello Coello [15]. By applying the Pareto ranking and keeping the historical records of the best solutions found by the particles ($pBest$ and $gBest$) as nondominated solutions generated in the past, PSO is a possible solution for MOPs. Figure 2.13 depicts the flowchart of a multiobjective PSO; further details are briefly described next.

The PSO for multiobjective optimization starts with a population of random solutions whose fitnesses are evaluated by the objective functions of the problem. Each particle flies through the problem space with a velocity, which is constantly updated by the particle's own experience and the best experience of its neighbors. In each iteration, the velocity and position of each particle are updated by the following equations:

$$v_i(g+1) = \omega v_i(g) + c_1 r_1 \left(pBest_i - pos_i^k \right) + c_2 r_2 \left(gBest - pos_i(g) \right) \tag{2.16}$$

$$pos_i(g+1) = pos_i(g) + v_i(g+1) \tag{2.17}$$

where $i = 1,2,\ldots,N$, ω is the inertia weight; c_1 and c_2 are the cognition factor and the social-learning factor, respectively; r_1 and r_2 are random numbers between 0 and 1; N is the population size; and $v_i(g)$ and $pos_i(g)$ are the velocity and position vectors at generation g, respectively.

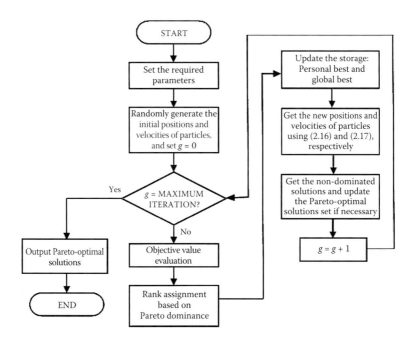

FIGURE 2.13
Flow chart of PSO.

2.2.9 Tabu Search

The TS was originally conceived in 1986 by Glover [35,36]. Since then, it has been successfully applied to combinatorial optimization problems due to its fast and aggressive search [1]. It is an optimization procedure that repeatedly moves from the current solution to the best in a list of neighboring solutions while avoiding being trapped in the local optima by referring to a tabu list of forbidden moves. Thus, it is considered as a local search technique guided by the use of adaptive or flexible memory structures [57].

When the neighbors of the current solution are searched, even if the neighbors are worse, the TS may still accept the new one under some probabilities so that better solutions can be explored. The visited best solution is stored in a memory pool, called the tabu list, so that searching backward is prohibited.

The traditional TS only works for single-objective optimal problems as the effort is geared toward trying to get the "best" solution among the neighbors. Therefore, a multiobjective TS (MOTS) for multiobjective optimization should work with a set of current solutions, which are simultaneously optimized toward the nondominated frontier [40].

The most direct way to implement MOTS is to expand the working object from a single solution to a population. Among the current population, a set of nondominated solutions is saved in the tabu list; thus, a local optimum

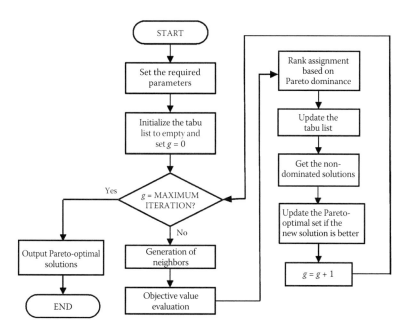

FIGURE 2.14
Flowchart of MOTS.

can be avoided as new neighbors can always be visited even if they are not nondominated solutions. The operations of MOTS is summarized by the flowchart depicted in Figure 2.14.

References

1. Aarts, E., Lenstra, J. K., *Local Search in Combinatorial Optimization*, New York: Wiley, 1997.
2. Anton, H., *Calculus with Analytic Geometry*, New York: Wiley, 1992.
3. Atmar, J. W., Speculation on the evolution of intelligence and its possible realization in machine form, Ph.D. dissertation, New Mexico State University, Las Cruces, 1976.
4. Bäck, T., *Evolutionary Algorithms in Theory and Practice: Evolution Strategies, Evolutionary Programming, Genetic Algorithms*, New York: Oxford University Press, 1996.
5. Bäck, T., Fogel, D. B., Michalewicz, Z. (Eds.), *Evolutionary Computation 1: Basic Algorithms and Operators*, Bristol, UK: Institute of Physics, 2000.
6. Bäck, T., Fogel, D. B., Michalewicz, Z. (Eds.), *Evolutionary Computation 2: Advanced Algorithms and Operators*, Bristol, UK: Institute of Physics, 2000.
7. Bäck, T., Hammel, U., Schwefel, H. P., Evolutionary computation: Comments on the history and current state, *IEEE Transactions on Evolutionary Computation*, 1(1), 3–17, 1997.

8. Beyer, H. G., *The Theory of Evolution Strategies*, Berlin: Springer, 2001.

9. Brassard, G., Bratley, P., *Fundamentals of Algorithmics*, Upper Saddle River, NJ: Prentice Hall, 1996.

10. Brassard, G., Bratley, P., *Algorithms: Theory and Practice*, Upper Saddle River, NJ: Prentice Hall, 1988.

11. Carter, M. W., Price, C. C., *Operations Research: A Practical Introduction*, Boca Raton, FL: CRC Press, 2001.

12. Coello Coello, C. A., Van Veldhuizen, D. A., Lamont, G. B., *Evolutionary Algorithms for Solving Multi-objective Problems*, New York: Kluwer, 2002.

13. Coello Coello, C. A., Toscano Pulido, G., A micro-genetic algorithm for multiobjective optimization, in *Proceedings of First International Conference on Evolutionary Multi-Criterion Optimization*, Zurich, March 2001, 126–140.

14. Coello Coello, C. A., Toscano Pulido, G., Multiobjective optimization using a micro-genetic algorithm, in *Proceedings of the Genetic and Evolutionary Computation Conference*, San Francisco, August 2001, 274–282.

15. Coello Coello, C. A., Pulido, G. T., Lechuga, M. S., Handling multiple objectives with particle swarm optimization, *IEEE Transactions on Evolutionary Computation*, 8(3), 256–279, 2004.

16. Colorni, A., Dorigo, M., Maniezzo, V., Distributed optimization by ant colonies, in *Proceedings of European Conference on Artificial Life*, Paris, December 1991, 134–142.

17. Colorni, A., Dorigo, M., Maniezzo, V., An investigation of some properties of an ant algorithm, in *Proceedings of the Conference on Parallel Problem Solving from Nature*, Brussels, September 1992, 509–520.

18. Cormen, T. H., Leiserson, C. E., Rivest, R. E., Stein, C., *Introduction to Algorithms*, Cambridge, MA: MIT Press, 2001.

19. Deb, K., *Multi-Objective Optimization Using Evolutionary Algorithms*, Chichester, UK: Wiley, 2001.

20. Deb, K., Agrawal, S., Pratap, A., Meyarivan, T., A fast elitist non-dominated sorting genetic algorithm for multi-objective optimization: NSGA-II, in *Proceedings of the Sixth International Conference on Parallel Problem Solving from Nature*, Paris, October 2000, 849–858.

21. Deb, K., Pratap, A., Agrawal, S., Meyarivan, T., A fast and elitist multiobjective genetic algorithm: NSGA-II. *IEEE Transactions on Evolutionary Computation*, 6(2), 182–197, 2002.

22. Dorigo, M., Blum, C., Ant colony optimization theory: A survey, *Theoretical Computer Science*, 344(2–3), 243–278, 2005.

23. Dorigo, M., Maniezzo, V., Colorni, A., The ant system: Optimization by a colony of cooperating agents, *IEEE Transactions on Systems, Man, and Cybernetics Part B: Cybernetics*, 26(2), 29–41, 1996.

24. Dumitrescu, D., Lazzerini, B., Jain, L. C., Dumitrescu, A., *Evolutionary Computation*, Boca Raton, FL: CRC Press, 2000.

25. Engelbrecht, A. P., *Computational Intelligence: An Introduction*, Chichester, UK: Wiley, 2002.

26. Erickson, M., Mayer, A., Horn, J., The niched Pareto genetic algorithm 2 applied to the design of groundwater remediation systems, in *Proceedings of the First International Conference on Evolutionary Multi-Criterion Optimization*, Zurich, March 2001, 681–695.

27. Faigle, U., Kern, W., Still, G., *Algorithmic Principles of Mathematical Programming*, Dordrecht, the Netherlands: Kluwer Academic, 2002.

28. Fogel, D. B., *Evolutionary Computation: Toward a New Philosophy of Machine Intelligence*, New York: IEEE Press, 2000.
29. Fogel, L. J., Owens, A. J., Walsh, M. J., *Artificial Intelligence through Simulated Evolution*, New York: Wiley, 1966.
30. Fogel, L. J., *Intelligence through Simulated Evolution: Forty Years of Evolutionary Programming*, New York: Wiley, 1999.
31. Fonseca, C. M., Fleming, P. J., Multiobjective optimization and multiple constraint handling with evolutionary algorithms—Part I: A unified formulation, *IEEE Transactions on System, Man and Cybernetic Part A: Systems and Humans*, 28(1), 26–37, 1998.
32. Fonseca, C. M., Fleming, P. J., Genetic algorithms for multiobjective optimization: Formulation, discussion and generalization, in *Proceedings of the Fifth International Conference on Genetic Algorithms*, Urbana-Champaign, IL, July 1993, 416–423.
33. Garey, M. R., Johnson, D. S., *Computers and Intractability: A Guide to the Theory of NP-completeness*, New York: Freeman, 1979.
34. Gentle, J. E., *Random Number Generation and Monte Carlo Methods*, New York: Springer, 2003.
35. Glover, F., Tabu search: Part I, *ORSA Journal on Computing*, 1(3), 190–206, 1989.
36. Glover, F., Tabu search: Part II, *ORSA Journal on Computing*, 2(1), 4–32, 1990.
37. Glover, F., Laguna, M., *Tabu Search*, Boston: Kluwer Academic, 1997.
38. Glover, F., Taillard, E., De Werra, D., A user's guide to tabu search, *Annals of Operations Research*, 41, 3–28, 1993.
39. Goldberg, D. E., *Genetic Algorithms in Search, Optimization and Machine Learning*, Reading, MA: Addison-Wesley, 1989.
40. Hansen, M. P., Tabu search for multiobjective optimization: MOTS, in *Proceedings of the 13th International Conference on Multiple Criteria Decision Making*, Cape Town, January 1997, 574–586.
41. Hillier, F. S., Lieberman, G. J., *Introduction to Mathematical Programming*, New York: McGraw-Hill, 1995.
42. Horn, J., Nafpliotis, N., Goldberg, D. E., A niched Pareto genetic algorithm for multiobjective optimization, in *Proceedings of the First International Conference on Evolutionary Computation*, Orlando, FL, June 1994, 1:82–87.
43. Horst, R., Tuy, H., *Global Optimization: Deterministic Approaches*, Berlin: Springer, 1996.
44. Husbands, P., Genetic algorithms in optimization and adaptation, in *Advances in Parallel Algorithms*, New York: Wiley, 1992, 227–276.
45. Johnsonbaugh, R., Schaefer, M., *Algorithms*, Upper Saddle River, NJ: Pearson Education, 2004.
46. Kennedy, J., Eberhart, R. C., Particle swarm optimization, in *Proceedings of the IEEE International Conference on Neural Networks*, Perth, 1995, 4:1942–1948.
47. Knowles, J. D., Corne, D. W., The Pareto archived evolution strategy: A new baseline algorithm for Pareto multiobjective optimization, in *Proceedings of Congress on Evolutionary Computation*, Washington, DC, July 1999, 1:98–105.
48. Knowles, J. D., Corne, D. W., Approximating the nondominated front using the Pareto archived evolution strategy, *Evolutionary Computation*, 8(2), 149–172, 2000.
49. Van Laarhoven, P. J. M., Aarts, E. H. L., *Simulated Annealing: Theory and Applications*, Dordrecht, the Netherlands: Reidel, 1987.
50. Lamont, G. B. (Ed.), *Compendium of Parallel Programs for the Intel iPSC Computers*, Dayton, OH: Department of Electrical and Computer Engineering, Air Force Institute of Technology, Wright-Patterson AFB, 1993.

51. Michalewicz, Z., Fogel, D. B., *How to Solve It: Modern Heuristics*, Berlin: Springer, 2000.

52. Michalewicz, Z., Michalewicz, M., Evolutionary computation techniques and their applications, in *Proceedings of IEEE International Conference on Intelligent Processing Systems*, Beijing, October 1997, 1:14–25.

53. Minoux, M., *Mathematical Programming: Theory and Algorithms*, Chichester, UK: Wiley, 1986.

54. Neapolitan, R., Naimipour, K., *Foundations of Algorithms*, Burlington, MA: Jones and Bartlett, 2009.

55. Oei, C. K., Goldberg, D. E., Chang, S. J., *Tournament Selection, Niching, and the Preservation of Diversity*, Technical Report (94002), Urbana: Illinois Genetic Algorithms Laboratory, University of Illinois, 1994.

56. Pearl, J., *Heuristics: Intelligent Search Strategies for Computer Problem Solving*, Reading, MA: Addison-Wesley, 1984.

57. Pirlot, M., General local search methods, *European Journal of Operational Research*, 92(3), 493–511, 1996.

58. Rardin, R. L., *Optimization in Operations Research*, Upper Saddle River, NJ: Prentice Hall, 1998.

59. Robert, C. P., Casella, G., *Monte Carlo Statistical Methods*, New York: Springer, 1999.

60. Zomaya, A. Y., Natural and simulated annealing, *IEEE Computing in Science and Engineering*, 3(6), 97–99, 2001.

61. Schwefel, H. P., *Evolution and Optimum Seeking*, New York: Wiley, 1995.

62. Schwefel, H. P., *Numerical Optimization of Computer Models*, Chichester, UK: Wiley, 1981.

63. Silverman, B. W., *Density Estimation for Statistics and Data Analysis*, London: Chapman and Hall, 1986.

64. Spitzer, F., *Principles of Random Walk*, New York: Springer, 1976.

65. Srinivas, N., Deb, K., Multiobjective function optimization using nondominated sorting genetic algorithms, *Evolutionary Computation*, 2(3), 221–248, 1994.

66. Van Veldhuizen, D.A., Lamont, G. B., Multiobjective evolutionary algorithms: Analyzing the state-of-the-art, *Evolutionary Computation*, 8(2), 125–147, 2000.

67. Yao, X., Global optimisation by evolutionary algorithms, in *Proceedings of IEEE International Symposium on Parallel Algorithms/Architecture Synthesis*, Fukushima, March 1997, 282–291.

68. Zabinsky, Z. B., *Stochastic Adaptive Search for Global Optimization*, Boston: Kluwer Academic, 2003.

69. Zitzler, E., Laumanns, M., Thiele, L., *SPEA2: Improving the Strength Pareto Evolutionary Algorithm*, Technical Report (TIK-Report 103), Zurich: Swiss Federal Institute of Technology, 2001.

70. Zitzler, E., Thiele, L., Multiobjective evolutionary algorithms: A comparative case study and the strength Pareto approach, *IEEE Transactions on Evolutionary Computation*, 3(4), 257–271, 1999.

3

Jumping Gene Computational Approach

3.1 Biological Background

3.1.1 Biological Jumping Gene Transposition

The discovery of mobile genetic elements in chromosomes has a long history. It can probably date back to the 1940s, when the Nobel Laureate in Physiology or Medicine, Barbara McClintock, studied the maize chromosome. This was even earlier than when the double-helix structure of deoxyribonucleic acid (DNA) was suggested by J. D. Watson and F. Crick [62,63].

As indicated in McClintock's work [8,14,33,34], the maize chromosome has two parts dissociated from each other and always breaks at a certain location. It was later found that this phenomenon is caused by the presence of autonomous and the nonautonomous transposable elements, called the activator (Ac) and dissociator (Ds), respectively.

The Ds itself is stable. However, with the presence of its corresponding Ac, the Ds will move and consequently introduce the breaking of a chromosome, as shown in Figure 3.1. Such a gene transposition mechanism also results in the colorization of the maize kernels. The transposition of the Ds in different cells is random, and the coloring is closely related to the stage of seed development during dissociation and the number of Ac's presented.

In general, the transposable elements (called transposons or jumping genes, JGs) are DNA sequences that constitute a fraction of their host genome. They can move around the genome via the transposition processes and have been identified in various organisms, including prokaryotes (e.g., bacteria) [23] as well as eukaryotes (e.g., plants, animals) [15].

The JGs can be divided into three different classes based on their modes of transposition [5,16,57]. The class I JG transposes through the replication of the element itself. First, the element is transcribed into a ribonucleic acid (RNA) intermediate. An enzyme, which reverses the transcription process, reversely transcribes the RNA into DNA, which is then inserted into a new chromosomal position.

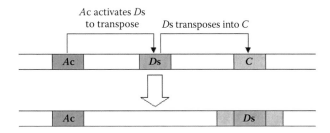

FIGURE 3.1
Transposition of Ds in a chromosome.

The class II JG moves directly from DNA to DNA through a conservative mechanism. In this mechanism, the excision of the donor JG is followed by its reinsertion elsewhere in the genome.

A class III JG shifts by replicating the element itself and inserting the copy into a new site. This operation is similar to that of a class I JG, but it does not involve transcription via an RNA intermediate.

According to the effects shown in Figure 3.2, the three classes of JGs can be simply categorized into two types: cut-and-paste JGs and copy-and-paste JGs. Table 3.1 tabulates some biological examples of these transposons.

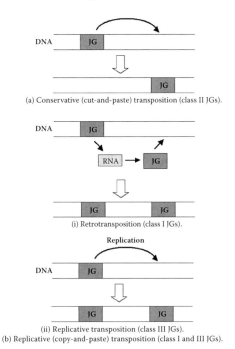

(a) Conservative (cut-and-paste) transposition (class II JGs).

(i) Retrotransposition (class I JGs).

(ii) Replicative transposition (class III JGs).
(b) Replicative (copy-and-paste) transposition (class I and III JGs).

FIGURE 3.2
Conservative (cut-and-paste) transposition and replicative (copy-and-paste) transposition.

TABLE 3.1

Examples of Cut-and-Paste and Copy-and-Paste Transposons

JGs	Examples	Organism
Copy-and-paste	Class III elements	
	(a) Replicative transposons	
	Tn3 elements	Bacteria
	Class I elements	
	(a) Retroviruslike elements (long terminal repeat [LTR] retrotransposons)	
	Ty1	Yeast
	Copia	*Drosophila*
	Gypsy	*Drosophila*
	(b) Retroposons	
	F, G, and I elements	*Drosophila*
	Telomere-specific retroposons (HeT-A, TART)	*Drosophila*
	LINES (e.g., L1)	Human
	SINES (e.g., Alu)	Human
Cut-and-paste	Class II elements	
	IS elements (e.g., IS50)	Bacteria
	Composite transposons (e.g., Tn5)	Bacteria
	Ac/Ds elements	Maize
	P elements	*Drosophila*
	Mariner elements	*Drosophila*
	Hobo elements	*Drosophila*
	Tc1 elements	Nematodes

3.1.2 Advantageous Effects of JG on Host Evolution

As revealed from various studies [2,4,35,36], the transposition of JGs affects the evolution of their hosts. In the past, transposable elements were regarded as selfish and junk genes that produced deleterious effects (e.g., fitness loss, genetic diseases, etc.) in their hosts. However, much evidence now confirms that these elements can indeed make significant contributions to new and beneficial host functions [3,20–22,25,26,29–31,37,39,44], and the following are some examples of their contributions:

1. The insertion of the transposon *P* element in the third intron of the *methuselah (mth)* gene was observed in *Drosophila melanogaster* [33]. Its presence gave the flies about a 35% increased average life span and higher resistance to different forms of stress, including starvation, high temperature, and the like.

2. The insertion of the retrotransposon Ty1 at the break site of the *mating type* (*MAT*) locus was observed in yeast [40,59]. It repaired

Homothallic switching (*HO*) endonuclease-induced double-strand breaks, which are lethal.

3. The *gag* region of the retrovirus of the human endogenous retrovirus-like (*HERV-L*) family derived the *Friend virus susceptibility 1* (*Fv1*) gene of mice [1]. It protected the host by conferring retroviral resistance.

4. Viviparous organisms with active immune systems face a problem that the maternal immune system treats the developing embryo as a "foreign" organism. The embryo would be attacked by the immune system if unprotected. However, endogenous retroelements in all mammals (e.g., *endogenous retrovirus-3* [*ERV3*] in humans) were expressed at high levels in specific tissues. The purpose of these tissue expressions (e.g., in the placenta) was to protect the embryo from being attacked by the maternal immune system [27,60].

3.2 Overview of Computational Gene Transposition

Inspired by the mechanisms of gene transposition, several related computational algorithms have been proposed, and they are briefly summarized next.

3.2.1 Sexual or Asexual Transposition

In References 47–51, gene transposition operations were suggested. The genes to be transposed are determined by referring to a flanking sequence that is randomly selected. An example is given in Figure 3.3. A flanking sequence consisting of two genes bc is randomly selected and indicated by the pointer p. It is then to search the next identical or inverse flanking sequence in the chromosome. If successfully found, a transposon is specified by including the genes starting at the end of the first flanking sequence to the end of the second flanking sequence. In the example given in Figure 3.3, dbc is the transposon.

The transposon can undergo sexual or asexual transposition. In sexual transposition, two chromosomes are involved, while only one chromosome is involved in asexual transposition. For sexual transposition, two different ways are possible to generate the resultant chromosomes, as given in Figure 3.3. The first method is called tournament-based transposition, for which the transposon is pasted at the position right after the flanking sequence found in another chromosome, but the original chromosome is kept unchanged. Another way is simple transposition. The two gene segments in the corresponding chromosomes are swapped, as shown in Figure 3.3.

For asexual transposition, after identifying the transposon by finding the first and the second flanking sequences, a third flanking sequence in the same chromosome is searched. An insertion point is then defined, as

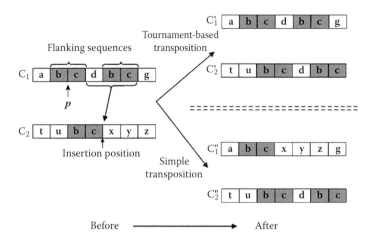

FIGURE 3.3
Example of tournament-based transposition and simple transposition for sexual transposition.

shown in Figure 3.4. The transposon is then cut from its original position and inserted in the space right after the third flanking sequence, where the space is spared by shifting some genes.

This gene transposition operation is suggested to replace crossover and serve as the major operation in a genetic algorithm (GA). By testing with 18 test functions taken from References 10, 17, 24, 38, and 61, it was reported in Simoes and Costa [48,49] that it outperforms a GA with crossover. These test functions cover various functional characteristics, including continuity/discontinuity, unimodal/multimodal, high/low/scalable dimensionality, stochastic/deterministic, quadratic/nonquadratic, and convex/nonconvex. It was concluded that, by increasing the diversity in the population using transposition, even with a small population of individuals, the results in most of the studied cases are much better than one-point, two-point, and uniform crossover. If the population size is large, transposition outperforms all types of crossover operations in all cases.

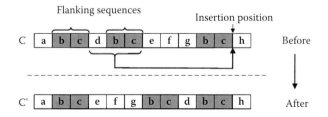

FIGURE 3.4
Example of asexual transposition.

However, note that the performance of transposition will be affected by the size of the flanking sequence, especially when a small population size is encountered. If a long sequence is defined, the average length of the transposon increases. The occurrence probability of transposition will then be reduced since the next same flanking sequence (or reverse flanking sequence) is less likely to be found. As a consequence of having fewer transpositions of genes, minimal effect is expected. Moreover, extra computational effort is required because the sequence searching is tedious on a bit-by-bit basis.

3.2.2 Bacterial Operations

Horizontal transmission is not the only cause of genetic transfer in chromosomes. Similar effects are also observed when bacteria attack a cell. In the past few years, a series of computational bacterial operations has been proposed. These operations, though inspired by another source, have somehow similar effects in altering a chromosome.

3.2.2.1 Transduction

Transduction [18,41–43,64] is a process involving a virus that accidentally picks up a copy of genes from a host cell and inserts it in the chromosome of an infected cell. With such a process, features of a single bacterium can be passed to the entire population.

The computational transduction mechanism is depicted in Figure 3.5. A chromosome is divided into p parts, each of which may represent a system parameter. The chromosome is then reproduced into m clones. A randomly chosen gene portion, say the ith part of $(m - 1)$ clones, is mutated. The elite among the m clones is then selected, and its ith part is transferred to the other $(m - 1)$ clones, replacing the ith part of all the other clones. The process of mutation-evaluation-selection-replacement is repeated, and each time a new part (i.e., not selected before) is randomly chosen. When all the p parts have been processed, only one clone is kept in the population; the others are discarded.

In References 18, 41–43, and 64, transduction is embedded into a GA to form a new algorithm, named the pseudobacterial genetic algorithm (PBGA). In a PBGA, transduction is performed prior to crossover, while mutation is abandoned. It has been used to solve several fuzzy system design problems, and it outperforms conventional GA by locating the optimal solutions more efficiently.

Transduction realizes a local improvement mechanism in parts in a chromosome while interactions among chromosomes are performed through the crossover operation. The performance of transduction highly depends on the choice of parts in the chromosomes, implying that the coding scheme is important. For example, each part of the chromosome represents a fuzzy

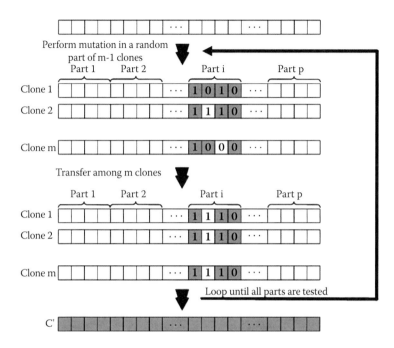

FIGURE 3.5
Example of transduction.

rule in fuzzy system design problems. If canonical binary encoding is used, transduction may not be an effective approach.

3.2.2.2 Conjugation

Conjugation [19,55,56] is another genetic operator inspired by bacterial recombination. It involves a unidirectional transfer of genetic material by direct cellular contact between a donor bacterial cell and a recipient cell. A kind of conjugation based on tournament selection was suggested [23,59]. Parents are randomly selected; then, the winner of the tournament becomes the donor, and the loser becomes the recipient.

Figure 3.6 depicts a simple example of conjugation. First, two points are randomly selected in the donor chromosome to identify the region for conjugation. The genes within the conjugation region are then transferred to the same region of the recipient.

Apparently, the same effect shown in Figure 3.6 can also be achieved by double-point crossover. However, there exists a significant difference between conjugation and crossover. The conjugation operator can be applied repeatedly by different donors onto the same recipient. Such a multiple conjugation is applicable where recombination takes place directly on the phenotype [55]. A study on the efficiency of conjugation for a practical design problem can be found in Perales-Gravan and Lahoz-Beltra [45].

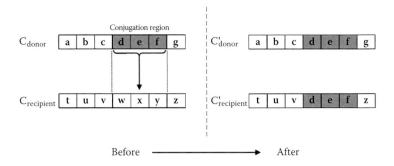

FIGURE 3.6
Example of conjugation.

3.2.2.3 Transformation

Transformation [52–54] mimics the biological capability of some bacteria to absorb fragments of DNA from the environment and then to reintegrate these gene segments in the recipient cells. An example is shown in Figure 3.7. A chromosome is selected from the mating pool; a gene segment is chosen from the gene segment pool. The segment is incorporated in the chromosome by replacing the genes after a randomly generated transformation point. In each generation, the old population contributes part of the new segment pool; the rest is created randomly.

Transformation is designed to replace crossover in the GA so that a higher genetic variation is provided. As studied [52,53], a better solution was found for several highly multimodal test functions as compared with the use of the standard GA. This was mainly because this new genetic operator can better preserve the diversity in the population of an evolutionary algorithm. Some preferable parameter settings for transformation have been suggested [55]. It was concluded that if the segment size increases, the results worsen since larger segments introduce more disruption in the individuals of the population.

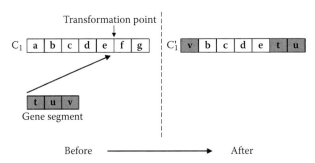

FIGURE 3.7
Example of transformation.

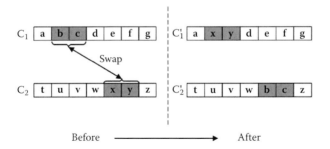

FIGURE 3.8
Example of translocation.

The major difference between transformation and other bacterial operations is that genetic diversity is provided from the environment rather than from the population itself. If crossover is replaced by transformation, a high level of diversity will be obtained. As the chromosomes do not have the initiative to pass on their gene segments, some existing important genes may be easily lost.

3.2.3 Other Operations

There also exist a few other operations that perform similar horizontal transmission without any biological inspiration. For example, a gene-swapping operation, called translocation, has been proposed [9] by which chromosomal segments are moved from one location to another. An example of translocation is depicted in Figure 3.8, which shows the swapping of two gene segments between two chromosomes. The positions of the segments are randomly selected, while the segment length is predefined.

3.3 Jumping Gene Genetic Algorithms

As mentioned in Section 3.1.2, biological gene transposition is able to provide new and advantageous host functions. In the same analogy, a computational transposition operation can be designed for a GA-based multiobjective evolutionary algorithm (MOEA) to enhance the search for novel as well as superior solutions. Thus, we proposed imitating and transcribing the biological transposition into a new computational genetic operation, the JG transposition [6,7,32,58]. Inspired by the biological JG as discussed in Section 3.1.1, two kinds of transposition, cut-and-paste and copy-and-paste, are proposed.

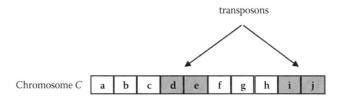

FIGURE 3.9
An example of a chromosome containing transposons.

3.3.1 Transposons in Chromosomes

Some consecutive genes in a chromosome can be selected as a transposon. Figure 3.9 depicts an example of two transposons, each of length two bits, randomly selected in a chromosome C. Both the number and the length of transposons in each chromosome can be greater than one but are user predefined.

The genes *a* to *j* in Figure 3.9 can be in any data form, such as binary, integer, or floating-point numbers, depending on the the chromosome encoding scheme in use. In some cases, for heterogeneous genes, say, integral and floating-point genes may be possible and coexist in a chromosome. For example, genes *d* and *e* form an integral transposon, while genes *i* and *j* give a floating-point transposon.

To avoid gene-type violation, transposition of genes should only be performed within the same data type. For example, if the required gene type at the locations of genes *d*, *e*, and *g* is integral but the one at the location of gene *f* is floating point, the integral transposon (i.e., genes *d* and *e*) is not allowed to jump to the positions of genes *f* and *g*. Otherwise, an invalid chromosome will be formed.

3.3.2 Cut-and-Paste and Copy-and-Paste Operations

As mentioned in Section 3.1, transposons can horizontally jump from one position to another position by cut-and-paste or copy-and-paste operation. The operations are given in Figures 3.10 and 3.11, respectively.

For the cut-and-paste operation, a transposon is cut from the original position and pasted into a new position of a chromosome. For the copy-and-paste operation, a transposon is replicated, and its copy is pasted into a new location of a chromosome, while the original one remains unchanged. In either case, the operations can be performed on the same chromosome or two different chromosomes, as shown in Figures 3.10 and 3.11.

The flowcharts of the cut-and-paste and copy-and-paste operations are shown in Figures 3.12 and 3.13, respectively. The operation made within the same chromosome or to a different chromosome is randomly selected, and there is no restriction regarding chromosome choice.

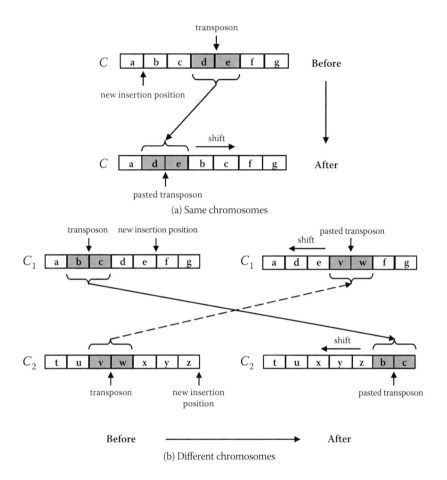

FIGURE 3.10
Cut-and-paste operation. (From Man, K. F., Tang, K. S., Kwong, S., Jumping-genes in evolutionary computing, in *Proceedings 30th Annual Conference of the IEEE Industrial Electronics Society*, 2004, Busan, Korea, 1268–1272.)

3.3.3 Jumping Gene Transposition

Referring to transposition in biology, the cut-and-paste and copy-and-paste operations are randomly chosen. As natural selection tends to be opportunistic but not prescient, the transposition processes are neither streamlined nor planned in advance. It turns out that transposition via either operation is similar to other genetic operations (i.e., crossover and mutation) that work based on the probability governed by the jumping rate p_j.

This can be easily implemented as follows: A random number r ranging between 0 and 1 is generated for each transposon. If $r \leq p_j$, the transposon will jump to another randomly generated valid position through the

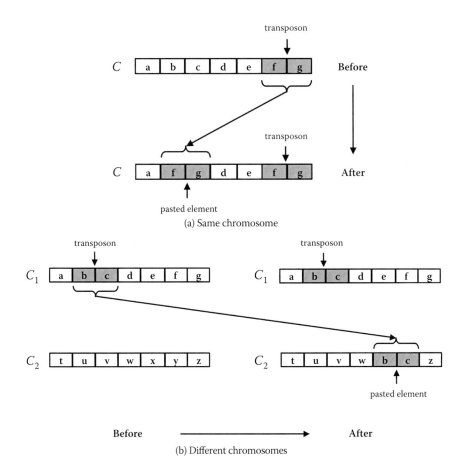

FIGURE 3.11
Copy-and-paste operation. (From Man, K. F., Tang, K. S., Kwong, S., Jumping-genes in evolutionary computing, in *Proceedings 30th Annual Conference of the IEEE Industrial Electronics Society*, 2004, Busan, Korea, 1268–1272.)

cut-and-paste or copy-and-paste operation. The overall flowchart of transposition is given in Figure 3.14.

The two JG operations are to be applied after parent selection but before crossover, as shown in Figure 3.15.

Therefore, this maintains the integrity and the basic flow of the original GA with the least modification. It was also demonstrated [6] that it outperforms the conventional GA for a large set of constrained and unconstrained multiobjective optimization problems, for which details are presented in Chapter 5.

3.3.4 Some Remarks

As compared with other operations described in Section 3.2, the described JG operational approach is much more general. For example, conjugation and

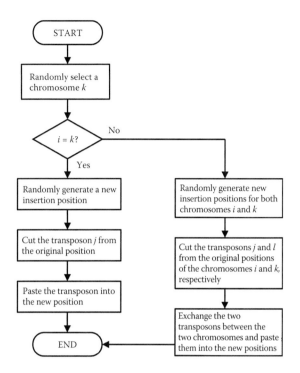

FIGURE 3.12
Flowchart of the cut-and-paste operation. (From Chan, T. M., Man, K. F., Kwong, S., Tang, K. S., A jumping gene paradigm for evolutionary multiobjective optimization, *IEEE Transactions on Evolutionary Computation*, 12(2), 143–159, 2008.)

translocation can simply be treated as special cases of copy-and-paste and cut-and-paste transpositions, respectively. Compared with Simoes's definition of a transposon, all genes have an equal chance to be selected in our JG approach without bias. In Simoes's definition, the length and the location of the transposon are determined by the presence of flanking sequences, while they can be any values predefined or randomly chosen under our proposed JG framework. Although Simoes's flanking sequences are randomly selected, because the transposon is defined as the segment between two identical or reverse flanking sequences, some genes will never become a transposon to pass on their genetic information if such identical or reverse sequence cannot be found.

3.4 Real-Coding Jumping Operations

As mentioned in Section 3.3.1, the transposon can be of any data type, and the designed JG operations can also be dealt with a GA using different coding schemes. However, precautions should be taken for a nonbinary type of

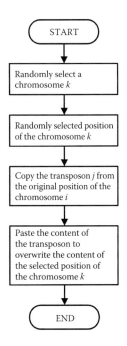

FIGURE 3.13
Flowchart of the copy-and-paste operation. (From Chan, T. M., Man, K. F., Kwong, S., Tang, K. S., A jumping gene paradigm for evolutionary multiobjective optimization, *IEEE Transactions on Evolutionary Computation*, 12(2), 143–159, 2008.)

transposon. In some cases, each gene may have its own bounded feasible range. A direct implementation of a cut-and-paste or copy-and-paste operation may result in an invalid chromosome type. A simple solution for this is to perform normalization of the value so that the boundness of the gene can be guaranteed.

An alternative method has also been suggested [46]. Instead of following the normal procedural design, other kinds of operations are used to attain transposition. The basic idea is similar to applying multiple crossover/mutation operators to enhance the population diversity. Figure 3.16 depicts the flowchart of this real-value jumping operation. Unlike the previous JG operations, there is no scope to differentiate between cut-and-paste and copy-and-paste.

Referring to Figure 3.16, the jumping operations are different when applied to a single chromosome or a pair of them. When a single chromosome undergoes the jumping operation, the following polynomial mutation operation [13] is applied:

$$y_i = x_i + (x_i^U - x_i^L)\delta_i \qquad (3.1)$$

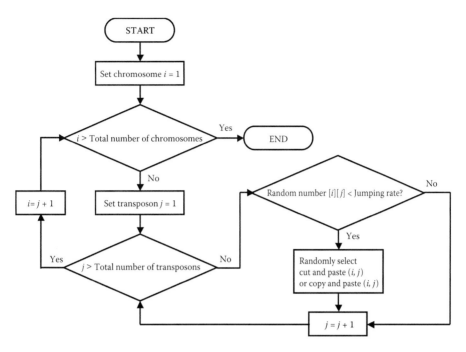

FIGURE 3.14
Flowchart of transposition. (From Chan, T. M., Man, K. F., Kwong, S., Tang, K. S., A jumping gene paradigm for evolutionary multiobjective optimization, *IEEE Transactions on Evolutionary Computation*, 12(2), 143–159, 2008.)

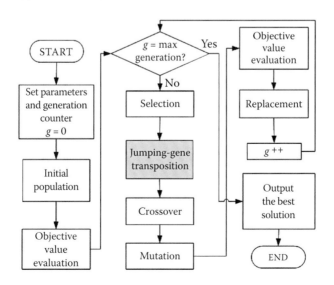

FIGURE 3.15
Genetic cycle of the JG GA. (From Tang, K. S., Kwong, S., Man, K. F., A jumping genes paradigm: Theory, verification and applications, *IEEE Circuits and Systems Magazine*, 8(4), 18–36, 2008.)

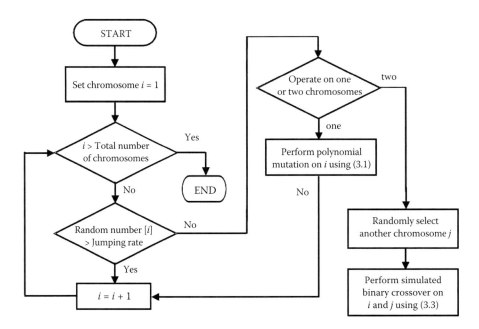

FIGURE 3.16
Flowchart of the real-coded jumping operations. (From Chan, T. M., Man, K. F., Kwong, S., Tang, K. S., A jumping gene paradigm for evolutionary multiobjective optimization, *IEEE Transactions on Evolutionary Computation*, 12(2), 143–159, 2008.)

where x_i is the gene to be mutated, y_i is the resultant gene, x_i^L and x_i^U are the lower and upper bounds of x_i, respectively, and

$$\delta_i = \begin{cases} (2r_i)^{1/(\eta_m+1)} - 1 & \text{if } r_i < 0.5 \\ 1 - [2(1-r_i)]^{1/(\eta_m+1)} & \text{if } r_i \geq 0.5 \end{cases} \tag{3.2}$$

with a random number $r_i \in [0,1]$ and a predefined value η_m. Similar to conventional mutation, it is a genewise operation, meaning that each gene may be mutated as specified in Equation (3.1) with a probability of an operational rate of, say, p_m.

In the case of two different chromosomes, the following simulated binary crossover [11,12] is performed as the jumping operation:

$$\begin{cases} H_1 = 0.5[(1-\beta)C_1 + (1+\beta)C_2] \\ H_2 = 0.5[(1+\beta)C_1 + (1-\beta)C_2] \end{cases} \tag{3.3}$$

where C_1 and C_2 are the two chromosomes selected for jumping operations, and H_1 and H_2 are the two offspring generated. The value of β is computed as

$$\beta = \begin{cases} (2a)^{1/(\eta_C+1)} & \text{if } a < 0.5 \\ [2(1-a)]^{-1/(\eta_C+1)} & \text{if } a \geq 0.5 \end{cases} \tag{3.4}$$

where $a \in [0,1]$ is a randomly generated value, and η_C is predefined.

Further details and performance comparisons can be found in Ripon, Kwong and Man [46].

References

1. Best, S., Tissier, P. L., Towers, G., Stoye, J. P., Positional cloning of the mouse retrovirus restriction gene Fv1, *Nature*, 382(6594), 826–829, 1996.
2. Bowen, N. J., Jordan, I. K., Transposable elements and the evolution of eukaryotic complexity, *Current Issues in Molecular Biology*, 4, 65–76, 2002.
3. Britten, R. J., DNA sequence insertion and evolutionary variation in gene regulation, *Proceedings of the National Academy of Sciences of the United States of America*, 93(18), 9374–9377, 1996.
4. Capy, P. (Ed.), *Evolution and Impact of Transposable Elements*, Dordrecht, the Netherlands: Kluwer Academic, 1997.
5. Capy, P., Bazin, C., Higuet, D., Langin, T., *Dynamics and Evolution of Transposable Elements*, Austin, TX: Landes Bioscience, 1998.
6. Chan, T. M., A generic jumping-gene paradigm: Concept, verification and applications, Ph.D. thesis, City University of Hong Kong, 2005.
7. Chan, T. M., Man, K. F., Kwong, S., Tang, K. S., A Jumping gene paradigm for evolutionary multiobjective optimization, *IEEE Transactions on Evolutionary Computation*, 12(2), 143–159, 2008.
8. Cohen, S. N., Shapiro, J. A., Transposable genetic elements, *Scientific American*, 242(2), 36–45, 1980.
9. De Falco, I., Iazzetta, A., Tarantino, E., Della Cioppa, A., On biologically inspired mutations: the translocation, in *Proceedings of Late Breaking Papers at the 2000 Genetic and Evolutionary Computation Conference*, Las Vegas, NV, July, 2000, 70–77.
10. De Jong, K. A., Analysis of the behavior of a class of genetic adaptive systems, Ph.D. dissertation, Department of Computer and Communication Science, University of Michigan, Ann Arbor, 1975.
11. Deb, K., Agarwal, S., Simulated binary crossover for continuous search space, *Complex Systems*, 9, 115–148, 1995.
12. Deb, K., Beyer, H., Self-adaptive genetic algorithms with simulated binary crossover, *Evolutionary Computation*, 9(2), 197–221, 2001.
13. Deb, K., Goyal, M., A combined genetic adaptive search (GeneAS) for engineering design, *Computer Science and Informatics*, 26(4), 30–45, 1996.
14. Fedoroff, N., Botstein, D. (Eds.), *The Dynamic Genome: Barbara McClintock's Ideas in the Century of Genetics*, New York: Cold Spring Harbor Laboratory Press, 1992.

15. Finnegan, D. J., Transposable elements in eukaryotes, *International Review of Cytology*, 93, 281–326, 1985.
16. Finnegan, D. J., Eukaryotic transposable elements and genome evolution, *Trends in Genetics*, 5, 103–107, 1989.
17. Fogel, D. B., *Evolutionary Computation: Toward a New Philosophy of Machine Intelligence*, Piscataway, NJ: IEEE Press, 1995.
18. Furuhashi, T., Nakaoka, K., Uchikawa, Y., A new approach to genetic based machine learning and an efficient finding of fuzzy rule, in *Proceedings of the IEEE/Nagoya-University World Wisepersons Workshop*, Nagoya, Japan, August 1994, 173–189.
19. Harvey, I., The microbial genetic algorithm, *Advances in Artificial Life, Lecture Notes in Computer Science 5778, Part II*, Berlin: Springer Verlag, 126–133, 2011.
20. Kidwell, M. G., Lisch, D. R., Transposable elements and host genome evolution, *Trends in Ecology and Evolution*, 15(3), 95–99, 2000.
21. Kidwell, M. G., Lisch, D. R., Transposable elements as sources of variation in animals and plants, *Proceedings of the National Academy of Sciences of the United States of America*, 94(15), 7704–7711, 1997.
22. Kidwell, M. G., Lisch, D. R., Perspective: Transposable elements, parasitic DNA, and genome evolution, *Evolution*, 55(1), 1–24, 2001.
23. Kleckner, N., Transposable elements in prokaryotes, *Annual Review of Genetics*, 15, 341–404, 1981.
24. Koon, G. H., Sebald, A. V., Some interesting test functions for evaluating evolutionary programming strategies, in *Proceedings of 4th Annual Conference on Evolutionary Programming*, San Diego, CA, February 1995, 479–499.
25. Kumar, A., Retrotransposons and their contributions to plant genome and gene evolution, in *Encyclopedia of Plant and Crop Science*, R. M. Goodman (Ed.), New York: Dekker, 2004, 1–5.
26. Labrador, M., Corces, V. G., Transposable element-host interactions: Regulation of insertion and excision, *Annual Review of Genetics*, 31, 381–404, 1997.
27. Larsson, E., Andersson, A. C., Nilsson, B. O., Expression of an endogenous retrovirus (ERV3 HERV-R) in human reproductive and embryonic tissues—Evidence for a function for envelope gene products, *Upsala Journal of Medical Sciences*, 99, 113–120, 1994.
28. Lin, Y. J., Seroude, L., Benzer, S., Extended life-span and stress resistance in the *Drosophila* mutant Methuselah, *Science*, 282(5390), 943–946, 1998.
29. Lorenc, A., Makalowski, W., Transposable elements and vertebrate protein diversity, *Genetica*, 118(2–3), 183–191, 2003.
30. Makalowski, W., Genomic scrap yard: How genomes utilize all that junk, *Gene*, 259(1–2), 61–67, 2000.
31. Makalowski, W., Not junk after all, *Science*, 300(5623), 1246–1247, 2003.
32. Man, K. F., Tang, K. S., Kwong, S., Jumping-genes in evolutionary computing, in *Proceedings 30th Annual Conference of the IEEE Industrial Electronics Society*, Busan, Korea, November 2004, 1268–1272.
33. McClintock, B., The origin and behavior of mutable loci in maize, *Proc. National Academy of Sciences USA*, 36, 344–355, 1950.
34. McClintock, B., Chromosome organization and genic expression, *Cold Spring Harbor Symposia on Quantitative Biology*, 16, 13–47, 1951.
35. McDonald, J. F. (Ed.), *Transposable Elements and Evolution*, Dordrecht, the Netherlands: Kluwer Academic, 1993.

36. McDonald, J. F. (Ed.), *Transposable Elements and Genome Evolution*, Dordrecht, the Netherlands: Kluwer Academic, 2000.

37. McDonald, J. F., Evolution and consequences of transposable elements, *Current Opinion in Genetics and Development*, 3(6), 855–864, 1993.

38. Michalewicz, Z., *Genetic Algorithms + Data Structures = Evolution Programs*, 3rd. ed. Berlin: Springer-Verlag, 1996.

39. Miller, W. J., McDonald, J. F., Pinsker, W., Molecular domestication of mobile elements, *Genetica*, 100(1–3), 261–270, 1997.

40. Moore, J. K., Haber, J. E., Capture of retrotransposon DNA at the sites of chromosomal double-strand breaks, *Nature*, 383(6601), 644–646, 1996.

41. Nawa, N., Furuhashi, T., Bacterial evolutionary algorithm for fuzzy system design, in *Proceedings of the IEEE International Conference on Systems, Man and Cybernetics*, San Diego, CA, October 1998, 3:2424–2429.

42. Nawa, N., Furuhashi, T., Hashiyama, T., Uchikawa, Y., A study of the discovery of relevant fuzzy rules using pseudo-bacterial genetic algorithm, *IEEE Transactions on Industrial Electronics*, 46(6), 1080–1089, 1999.

43. Nawa, N., Hashiyama, T., Furuhashi, T., Uchikawa, Y., A study on fuzzy rules using pseudo-bacterial genetic algorithm with adaptive operator, in *Proceedings of the IEEE International Conference on Evolutionary Computation*, Indianapolis, IN, April 1997, 589–593.

44. Nekrutenko, A., Li, W. H., Transposable elements are found in a large number of human protein-coding genes, *Trends in Genetics*, 17(11), 619–621, 2001.

45. Perales-Gravan, C., Lahoz-Beltra, R. An AM radio receiver designed with a genetic algorithm based on a bacterial conjugation genetic operator, *IEEE Transactions on Evolutionary Computation*, 12(2), 129–142, 2008.

46. Ripon, K. S. N., Kwong, S., Man, K. F., A real-coding jumping gene genetic algorithm for multiobjective optimization, *Information Sciences*, 177(2), 632–654, 2007.

47. Simoes, A., Costa, E., Transposition: A biologically inspired mechanism to use with genetic algorithms, in *Proceedings of the Fourth International Conference on Neural Networks and Genetic Algorithms*, Portoroz, Slovenia, April 1999, 178–186.

48. Simoes, A., Costa, E., Enhancing transposition performance, in *Proceedings of Congress on Evolutionary Computation*, Washington DC, July 1999, 1434–1441.

49. Simoes, A., Costa, E., Transposition versus crossover: An empirical study, in *Proceedings of the Genetic and Evolutionary Computation Conference*, Orlando, FL, July 1999, 612–619.

50. Simoes, A., Costa, E., Using genetic algorithms with sexual or asexual transposition: A comparative study, in *Proceedings of the Congress on Evolutionary Computation*, San Diego, CA, July 2000, 1196–1203.

51. Simoes, A., Costa, E., Using genetic algorithms with asexual transposition, in *Proceedings of the Genetic and Evolutionary Computation Conference*, Las Vegas, NV, July 2000, 323–330.

52. Simoes, A., Costa, E., Using biological inspiration to deal with dynamic environments, in *Proceedings of the Seventh International Conference on Soft Computing*, Brno, Czech Republic, June 2001, 7–12.

53. Simoes, A., Costa, E., On biologically inspired genetic operators: using transformation in the standard genetic algorithm, in *Proceedings of the Genetic and Evolutionary Computation Conference*, San Francisco, July 2001, 584–591.

54. Simoes, A., Costa, E., Parametric study to enhance genetic algorithm's performance when using transformation, in *Proceedings of the Genetic and Evolutionary Computation Conference*, New York, July 2002, 697.

55. Smith, P., Finding hard satisfiability problems using bacterial conjugation, in *Proceedings of AISB96 Workshop on Evolutionary Computing*, University of Sussex, Brighton, UK, April 1996, 236–244.

56. Smith, P., Conjugation–A bacterially inspired form of genetic recombination, in *Proceedings of Late Breaking Papers of the First Annual Conference on Genetic Programming*, Stanford, CT, July 1996, 167–176.

57. Snustad, D. P., Simmons, M. J., *Principles of Genetics*, New York: Wiley, 2003.

58. Tang, K. S., Kwong, S., Man, K. F., A jumping genes paradigm: Theory, verification and applications, *IEEE Circuits and Systems Magazine*, 8(4), 18–36, 2008.

59. Teng, S. C., Kim, B., Gabriel, A., Retrotransposon reverse-transcriptase-mediated repair of chromosomal breaks, *Nature*, 383(6601), 641–644, 1996.

60. Venables, P. J. W., Brookes, S. M., Griffiths, D., Weiss, R. A., Boyd, M. T., Abundance of an endogenous retroviral envelope protein in placental trophoblasts suggests a biological function, *Virology*, 211(2), 589–592, 1995.

61. Whitley, D., Mathias, K., Rana, S., Dzubera, J., Building better test functions, in *Proceedings of Sixth International Conference on Genetic Algorithms*, Pittsburgh, July 1995, 239–246.

62. Watson, J. D., *The double helix: A personal account of the discovery of the structure of DNA*, New York: Atheneum, 1968.

63. Watson, J. D., Crick, F. H. C., A structure of nucleic acids, *Nature*, 171(4356), 737–738, 1953.

64. Yoshikawa, T., Furuhashi, T., Uchikawa, Y., Knowledge acquisition of fuzzy control rules for mobile robots using DNA coding method and pseudo-bacterial genetic algorithm, in *Proceedings of First Asia-Pacific Conference on Simulated Evolution and Learning*, Taejon, South Korea, November 1996, 126–135.

4

Theoretical Analysis of Jumping Gene Operations

4.1 Overview of Schema Models

4.1.1 Schema

A schema ξ is a subset of the search space in which all the strings share a particular set of defined values. In the context of a genetic algorithm (GA) that operates with binary strings, a schema is a string of symbols taken from the set of three symbols {0, 1, *}, where 0 and 1 are called definite bits or actual bits, and the wild character * is interpreted as a don't-care bit that can be either 0 or 1. With the introduction of the don't-care bit, a single schema can represent several bit strings. A similar concept is also applicable for representations in other bases.

Take a simple example of three bits; a schema **1 represents four strings: 001, 011, 101, 111. This indicates a two-dimensional (2D) surface, as shown in Figure 4.1, while *** represents all the possible solutions or the three-dimensional (3D) cube.

To study a scheme, two of its attributes are of importance. One is called the order of a schema $o(\xi)$, representing the number of actual bits in the schema. The next is the defining length of a schema, denoted $L_d(\xi)$, defined as the distance between the leftmost and the rightmost actual bits. For example, $o(*0**10) = 3$ and $L_d(*0**10) = 4$. Further explanation of schema can be found in References 8 and 9.

4.1.2 Holland's Model

The most well-known mathematical model based on a schema must be the Holland schema theorem [9]. It is a typical approximate schema theorem, having the same level of coarse graining on both sides of the model equation. It predicts the expected number of strings belonging to a schema in a population being evolved from one generation to the next, under the operations

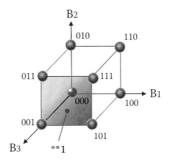

FIGURE 4.1
An example of a three-bit schemata. (From Tang, K. S., Yin, R. J., Kwong, S., Ng, K. T., Man, K. F., A theoretical development and analysis of jumping gene genetic algorithm, *IEEE Transactions on Industrial Informatics*, 7(3), 2011, 408–418.)

of selection, one-point crossover, and bit mutation. The schema growth equation is summarized as follows:

$$E[m(\xi, t+1)] \geq N \cdot p(\xi, t) \cdot (1 - p_m)^{o(\xi)} \cdot \left[1 - p_x \frac{L_d(\xi)}{L-1} \right] \qquad (4.1)$$

where p_m and p_x are the operational probabilities of mutation and crossover, respectively; L is the length of the chromosome; N is the population size; $E[m(\xi, t+1)]$ is the expected number of chromosomes matching with the schema ξ at generation $(t+1)$; and $p(\xi, t)$ is the probability of selecting the schema ξ as parent.

The term $(1 - p_m)^{o(\xi)}$ gives the probability that a schema survives after mutation. The term $p_x \frac{L_d(\xi)}{L-1}$ expresses the case that crossover is performed at the crossover point in between the actual bits. It thus approximates the destructive rate of a schema due to crossover.

Assuming that the fitness proportionate selection is used for selecting parents, it can be derived that

$$p(\xi, t) = \frac{m(\xi, t) f(\xi, t)}{N \bar{f}(t)} \qquad (4.2)$$

where $m(\xi, t)$ is the number of chromosomes belonging to schema ξ at generation t, $f(\xi, t)$ is the average fitness of the chromosomes belonging to ξ, and $\bar{f}(t)$ is the average fitness of the population.

The inequality (4.1) can only provide a lower bound on the expected number of chromosomes that belong to a specific schema in the next generation; hence it is an approximation. It is also pessimistic in the sense that it only considers the destruction of a schema, neglecting the possible reconstruction of the schema, which certainly plays an equally important role as destruction during evolution. Due to this shortcoming, the usefulness of schemata and the schema theorem has been widely criticized [1,6,7]. The main criticism is that the schema theorem cannot be practically used to predict the behavior of a GA over multiple generations. It may not even be meaningful since only a lower bound is presented.

However, in spite of many criticisms, it does not mean that the schema theorem is useless. Efforts have been made to have the schema theory be a more quantitative one. By using the microscopic dynamics of GA together with the appropriate measurement function [1], the schema theorem is expressed to show how changes in different macroscopic properties of populations in a GA can be derived. It indeed proves the intuitive idea about what makes a GA work; that is, offspring with above-average fitness can be produced by recombining schemata with above-average fitness.

4.1.3 Stephens and Waelbroeck's Model

To further improve Holland's model, Stephens and Waelbroeck derived a new schema evolution equation that gave an exact formulation (rather than a lower bound) for selection, single-point crossover, and mutation [13–15]. This theorem has been used as a starting point for many other results explaining the behavior of a GA over multiple generations, with the assumption of an infinite population size.

In Stephens and Waelbroeck's model, the effects of schema construction are made explicit. An exact formulation for the expected number of chromosomes belonging to a schema can then be obtained by considering both schema construction and destruction. In the process of the GA, schema construction commonly occurs, and it is important for the search. Offspring belonging to ξ can be formed by mating two chromosomes that do not belong to schema ξ. A similar effect can be observed by mutation.

It is possible to determine the exact number of chromosomes belonging to a schema in the next generation by considering all possible ways that schemata may survive, be destroyed, or be created. A major advantage of such an exact theorem is that the expected number of chromosomes matching a schema can be predicted over multiple generations, which clearly overcomes the shortcomings of Holland's theorem.

In the following, the exact schema evolution equation derived in References 13–15 is reviewed. The equation is derived for schemata evolving under the effects of three genetic operators: proportional selection, crossover, and mutation. For simplicity, single-point crossover is considered (refer to [12] for uniform crossover). Instead of calculating the expected number of chromosomes matching a schema, the expected relative proportion is considered as follows:

$$p(\xi,t) = \frac{m(\xi,t)}{N} \tag{4.3}$$

where $m(\xi,t)$ is the number of chromosomes matching the schema ξ at generation t, and N is the population size.

If mutation is carried out after crossover, the expected proportion of schema ξ in the population is

$$p(\xi, t+1) = P(\xi \to \xi) p_c(\xi, t) + \sum_{\bar{\xi}_i} P(\bar{\xi}_i \to \xi) p_c(\bar{\xi}_i, t) \tag{4.4}$$

where

$$P(\xi \to \xi) = (1 - p_m)^{o(\xi)} \quad \text{and} \quad P(\bar{\xi}_i \to \xi) = p_m^{d^H(\xi, \bar{\xi}_i)} (1 - p_m)^{o(\xi) - d^H(\xi, \bar{\xi}_i)} \tag{4.5}$$

The former is the probability that no mutation is carried out on schema ξ; the latter is the probability that a different schema $\bar{\xi}_i$ is mutated into schema ξ. $d^H(\xi, \bar{\xi}_i)$ is the Hamming distance between schemata ξ and $\bar{\xi}_i$.

The quantity $p_c(\xi, t)$ in Equation (4.4) is the expected proportion of chromosomes belonging to schema ξ after selection and single-point crossover. Explicitly,

$$p_c(\xi, t) = p'(\xi, t) - \frac{p_x}{L-1} \sum_{k=1}^{L-1} (p'(\xi, t) - p'(\xi_L(k), t) p'(\xi_R(k), t)) \tag{4.6}$$

where k is the crossover point, $\xi_L(k)$ is the schema obtained by replacing all the elements of ξ from position $k+1$ to position L with don't-care bits (*), and $\xi_R(k)$ is the schema obtained by replacing all the elements of ξ from position 1 to position k with don't-care bits.

Finally, $p'(\xi, t)$ in Equation (4.6) is the expected proportion of schema ξ after the proportional selection process, which is calculated as follows:

$$p'(\xi, t) = \frac{f(\xi, t)}{\bar{f}(t)} p(\xi, t) \tag{4.7}$$

Equations (4.4) and (4.6) take into account the effects of both destruction and construction of schema during mutation and crossover, respectively. They thus yield an exact expression for the expectation values of $m(\xi, t)$, that is, the number of strings matching ξ. In the limit of $N \to \infty$, Equations (4.4)–(4.7) present the correct probability distribution governing GA evolution.

It should also be noted that the development of an exact schema theorem is useful for providing insights on the effect of different operations. For example, as shown in Reference [12], mutation can achieve higher levels of disruption than uniform crossover, while uniform crossover is more powerful from a construction point of view.

4.2 Exact Schema Theorem for Jumping Gene Transposition

For a GA incorporating jumping gene (JG) operations, the exact schema formulae for selection, crossover, and mutation can be directly borrowed from the ones derived by Stephens and Waelbroeck, as described in Section 4.1.3. Therefore, our focus is to derive the schema evolution equations for the two JG transpositions, that is, the copy-and-paste and cut-and-paste operations [16,20].

4.2.1 Notations and Functional Definitions

Before describing the derivation of the equations, the notations and functions are given.

4.2.1.1 Notations

L	The length of the binary string (i.e., the chromosome length).
$S^{(i)}$	The set of all schemata with length i, and for simplicity, $S^{(L)} \equiv S$.
$V^{(i)}$	The superset of all sets formed by integers 0 to $(i-1)$, and for simplicity, $V^{(L)} \equiv V$. For example, if $L = 3$, $V^{(3)} = \{\phi, \{0\}, \{1\}, \{2\}, \{0,1\}, \{0,2\}, \{1,2\}, \{0,1,2\}\}$.
Z_{odd}, Z_{even}	The subsets of S, $Z_{odd} = \{\xi : zeros(\xi) = 2n+1\}$, $Z_{even} = \{\xi : zeros(\xi) = 2n\}$, for some integer n, and $zeros(\xi)$ returns the number of zeros in schema $\xi \in S$. It is also denoted that $Z'_{odd} = Z_{even}$ and $Z'_{even} = Z_{odd}$.
	For example, if $\xi = *01**0$, since $zeros(\xi) = 2$, we can get $\xi \in Z_{even}$.
$o(\xi)$	The order of schema ξ.
$L_d(\xi)$	The defining length of schema ξ.
L_g	The length of transposon, where $1 \le L_g < L$.

4.2.1.2 Functional Definitions

Definition 4.1

A map f_L is defined as $f_L : S \to V$, such that for $v \in V$, $v = f_L(\xi)$ returns the locations of all the actual bits in schema $\xi \in S$. It is also assumed that the location begins from 0. ∎

Definition 4.2

A map f_T is defined as $f_T : S \times V \to S^{(i)}$, such that for $\xi_2 \in S^{(i)}$, $\xi_2 = f_T(\xi_1, v)$ is formed by copying the bits from schema $\xi_1 \in S$ according to the locations specified in $v \in V$, where $i = size(v)$. ∎

Definition 4.3

The primary schemata competition set [3, 10, 19] of schema $\xi \in S$ is defined as

$$S_\xi = \{\xi_i \in S : f_L(\xi_i) = f_L(\xi)\} \qquad (4.8) \quad \blacksquare$$

Remark 4.1

If the order of schema ξ is $o(\xi) = n$, S_ξ is called the n-order primary schemata competition set. Any two schemata in S_ξ are said to have an n-order primary schemata competition relation. It is also denoted that $S_{\bar{\xi}} = S_\xi \setminus \{\xi\}$ contains all the competitors of schema ξ. $\blacksquare$

> **Example 4.1**
>
> Let $\xi = **0*0*$, then $f_L(\xi) = \{2, 4\}$ and $f_T(\xi, \{2, 3\}) = 0*$. We also have $S_\xi = \{**0*0*, **0*1*, **1*0*, **1*1*\}$, $S_{\bar{\xi}} = \{**0*1*, **1*0*, **1*1*\}$.

Definition 4.4

The bit distance between two bits x_i and y_i, where $x_i, y_i \in \{0, 1, *\} \equiv B$, is defined as

$$d(x_i, y_i) = \begin{cases} 1 & x_i = * \\ 1 & x_i \neq *, y_i \neq *, x_i = y_i \\ 0 & x_i \neq *, y_i \neq *, x_i \neq y_i \\ 0.5 & x_i \neq *, y_i = * \end{cases}$$

$$(4.9)$$

describing how close y_i matches x_i. $\blacksquare$

Remark 4.2

It should be noted that $d(x_i, y_i) \neq d(y_i, x_i)$. $\blacksquare$

Remark 4.3

For the case of $x_i \neq *$ and $y_i = *$, $d(x_i, y_i) = 0.5$. It assumes that a don't-care bit has an equal chance to be 0 or 1. $\blacksquare$

Definition 4.5

The regional similarity of two strings v_1 and v_2, where $v_i, v_j \in B^m$, is specified by

$$\delta(v_1, v_2) = \prod_{j=0}^{m-1} d(v_1(j), v_2(j))$$

$$(4.10)$$

where $v_i(j)$ is the jth bit of v_i. Note that $\delta(v_1, v_2)$ ranges from 0 to 1. $\blacksquare$

Remark 4.4

To calculate the regional similarity of ξ_1 in region V_1 and ξ_2 in region V_2, let $v_1 = f_T(\xi_1, V_1)$ and $v_2 = f_T(\xi_2, V_2)$; then the regional similarity can be calculated as

$$\delta(f_T(\xi_1, V_1), f_T(\xi_2, V_2)) = \delta(v_1, v_2) = \prod_{j=0}^{\tilde{m}-1} d(v_1(j), v_2(j)) \qquad (4.11)$$

where $\tilde{m}$ is the size of region V_1 and V_2 (the size of both regions must be the same). For simplicity,

$$\delta(f_T(\xi_1, V_1), f_T(\xi_2, V_2)) \equiv \Delta(\xi_1, V_1; \xi_2, V_2) \qquad (4.12) \qquad \blacksquare$$

Example 4.2

Consider two schemata $\xi_1 = {}^*01{}^*1{}^{**}$ and $\xi_2 = 0{}^{**}1{}^*01$ and two regions with $V_1 = [2, 4]$ and $V_2 = [4, 6]$; the regional similarity of ξ_1 in V_1 and ξ_2 in V_2 is calculated as

$$\Delta(\xi_1, V_1; \xi_2, V_2) = \delta(1{}^*1, \ \ {}^*01) = d(1, {}^*) \times d({}^*, 0) \times d(1, 1) = 0.5 \times 1 \times 1 = 0.5 \cdot$$

4.2.2 Exact Schema Evolution Equation for Copy-and-Paste

Now consider a primary schemata competition set S_ξ with $o(\xi_i) = n$, $\xi_i \in S_\xi$, and hence $size(S_\xi) = 2^n \equiv M$. The defining length of schema ξ_i is $L_d = p_2 - p_1$, with $p_1 = \min(f_L(\xi_i))$ and $p_2 = \max(f_L(\xi_i))$ denoting the bit locations of the first and last actual bits, respectively. It is assumed that a copy-and-paste operation is performed on schemata $\xi_m, \xi_n \in S_\xi$. A transposon of length L_g is copied from ξ_m at position c and pasted into position k of ξ_n, and the resultant schema is denoted as ξ_n'.

Figure 4.2 depicts the operation for which some important notations are given as follows:

1. **Transposon region**: $G_{c,k} = [c + \max(p_1 - k, 0), c + \min(p_1 + L_d - k, L_g - 1)]$ specifies the bit locations of selected transposons in ξ_m, which fall into the defining length region $[p_1, p_2]$ of the targeted schema ξ, after pasting;
2. **Pasted region**: $V_k = [\max(p_1, k), \min(p_1 + L_d, k + L_g - 1)]$ specifies the bit locations within the region $[p_1, p_2]$ of ξ where the transposon is pasted;
3. **Unchanged region**: $V_k' = [p_1, \min(p_1 + L_d, k - 1)] \cup [\max(p_1, k + L_g), p_1 + L_d]$ specifies the bit locations of ξ within the region $[p_1, p_2]$ but not belonging to V_k.

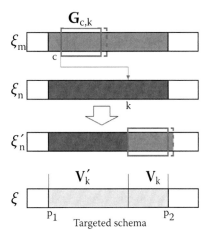

FIGURE 4.2
Copy-and-paste on schemata ξ_m and ξ_n, where ξ'_n is the resultant schema that will be compared with targeted schema ξ. (From Tang, K. S., Yin, R. J., Kwong, S., Ng, K. T., Man, K. F., A theoretical development and analysis of jumping gene genetic algorithm, *IEEE Transactions on Industrial Informatics*, 7(3), 2011, 408–418.)

The population size is assumed to be infinite, and instead of calculating the exact number of strings belonging to some schemata, the expected occurrence probability $P(\xi_i, t)$, $\xi_i \in S_\xi$ is the focus. The derivation of the schema evolution equation follows the exact formulations of other GA dynamics, in which the effects of both destruction and construction are considered.

Depending on the schemata ξ_m and ξ_n, four possible cases are obtained.

Case A

$\xi_m = \xi$ and $\xi_n = \xi$. (A transposon is copied from schema ξ and pasted onto schema ξ.)

If the resultant schema ξ'_n no longer belongs to ξ, that is, $\xi'_n \neq \xi$, after the transposition, schema ξ is hence destroyed, and the condition is

$$\Delta(\xi, V_k; \xi'_n, V_k) \neq 1 \Rightarrow \Delta(\xi, V_k; \xi, G_{c,k}) \neq 1 \tag{4.13}$$

By considering all the possible combinations of c and k, the destructive rate PD_A of ξ is then computed as

$$PD_A = \frac{p_{copy}}{(L - L_g + 1)^2} \sum_{c=0}^{L-L_g} \sum_{k=0}^{L-L_g} [1 - \Delta(\xi, V_k; \xi, G_{c,k})] \tag{4.14}$$

where p_{copy} is the operational rate of the copy-and-paste operation, and the factor $\frac{1}{(L-L_g+1)^2}$ indicates the probability of selecting a particular pair of c and k. The occurrence probability of case A is given as

$$E_A(t) = P(\xi,t)\,P(\xi,t) \tag{4.15}$$

Hence, the expected destruction of ξ in this case is calculated as

$$P_A(\xi,t) = PD_A \times E_A(t)$$

$$= \frac{p_{copy}}{(L-L_g+1)^2} \sum_{c=0}^{L-L_g} \sum_{k=0}^{L-L_g} [1-\Delta(\xi,V_k;\xi,G_{c,k})]\,P(\xi,t)P(\xi,t) \tag{4.16}$$

Case B

$\xi_m = \xi$ and $\xi_n \in S_{\bar{\xi}}$. (A transposon is copied from schema ξ and pasted onto another schema ξ_n where $\xi_n \neq \xi$.)

If $\xi'_n = \xi$ after transposition, schema ξ is constructed. This can be described by the regional similarity expressed as

$$\Delta(\xi,V_k;\xi'_n,V_k)\Delta(\xi,V'_k;\xi'_n,V'_k) \neq 0 \Rightarrow \Delta(\xi,V_k;\xi,G_{c,k})\Delta(\xi,V'_k;\xi_n,V'_k) \neq 0. \tag{4.17}$$

The constructive rate PC_B of ξ from ξ_n is then obtained by

$$PC_B = \frac{p_{copy}}{(L-L_g+1)^2} \sum_{c=0}^{L-L_g} \sum_{k=0}^{L-L_g} [\Delta(\xi,V_k;\xi,G_{c,k})\Delta(\xi,V'_k;\xi_n,V'_k)] \tag{4.18}$$

while the selection probability of a particular pair of ξ_m and ξ_n is

$$E_B(t) = P(\xi,t)P(\xi_n,t) \tag{4.19}$$

Hence the expected construction of ξ in this case is

$$P_B(\xi,t) = \sum_{\xi_n \in S_{\bar{\xi}}} PC_B \times E_B(t)$$

$$= \frac{p_{copy}}{(L-L_g+1)^2}$$

$$\times \sum_{\xi_n \in S_{\bar{\xi}}} \sum_{c=0}^{L-L_g} \sum_{k=0}^{L-L_g} [\Delta(\xi,V_k;\xi,G_{c,k})\Delta(\xi,V'_k;\xi_n,V'_k)]\,P(\xi,t)P(\xi_n,t) \tag{4.20}$$

Case C

$\xi_m \in S_{\bar{\xi}}$ and $\xi_n = \xi$. (A transposon is copied from schema ξ_m, where $\xi_m \neq \xi$, and pasted onto schema ξ.)

If $\xi'_n \neq \xi$ after transposition, schema ξ is destroyed. Similar to case A, the condition of regional similarity is

$$\Delta(\xi, V_k; \xi'_n, V_k) \neq 1 \Rightarrow \Delta(\xi, V_k; \xi_m, G_{c,k}) \neq 1 \tag{4.21}$$

The destructive rate PD_C is then derived as

$$PD_C = \frac{p_{copy}}{(L - L_g + 1)^2} \sum_{c=0}^{L-L_g} \sum_{k=0}^{L-L_g} [1 - \Delta(\xi, V_k; \xi_m, G_{c,k})] \tag{4.22}$$

while the probability of a particular pair of ξ_m and ξ_n being selected is given by

$$E_C(t) = P(\xi_m, t)P(\xi, t) \tag{4.23}$$

The expected destruction of ξ is

$$P_C(\xi, t) = \sum_{\xi_m \in S_{\bar{\xi}}} PD_C \times E_C(t)$$

$$= \frac{p_{copy}}{(L - L_g + 1)^2} \sum_{\xi_m \in S_{\bar{\xi}}} \sum_{c=0}^{L-L_g} \sum_{k=0}^{L-L_g} [1 - \Delta(\xi, V_k; \xi_m, G_{c,k})] \, P(\xi_m, t)P(\xi, t) \tag{4.24}$$

Case D

$\xi_m, \xi_n \in S_{\bar{\xi}}$. (A transposon is copied from schema ξ_m and pasted onto another schema ξ_n, where both of them are not ξ.)

To have $\xi'_n = \xi$ after transposition, that is, schema ξ is constructed from ξ_n, similar to case B, the following condition is obtained:

$$\Delta(\xi, V_k; \xi'_n, V_k)\Delta(\xi, V'_k; \xi'_n, V'_k) \neq 0 \Rightarrow \Delta(\xi, V_k; \xi_m, G_{c,k})\Delta(\xi, V'_k; \xi_n, V'_k) \neq 0 \tag{4.25}$$

The constructive rate PC_D for ξ in this case is hence derived as

$$PC_D = \frac{p_{copy}}{(L - L_g + 1)^2} \sum_{c=0}^{L-L_g} \sum_{k=0}^{L-L_g} [\Delta(\xi, V_k; \xi_m, G_{c,k})\Delta(\xi, V'_k; \xi_n, V'_k)] \tag{4.26}$$

The probability of a particular pair of ξ_m and ξ_n is

$$E_D(t) = P(\xi_m, t)P(\xi_n, t) \tag{4.27}$$

and the expected construction of ξ in this case is computed as

$$P_D(\xi, t) = \sum_{\xi_m \in S_{\bar{\xi}}} \sum_{\xi_n \in S_{\bar{\xi}}} PC_D \times E_D(t)$$

$$= \frac{p_{copy}}{(L - L_g + 1)^2} \tag{4.28}$$

$$\times \sum_{\xi_m \in S_{\bar{\xi}}} \sum_{\xi_n \in S_{\bar{\xi}}} \sum_{c=0}^{L-L_g} \sum_{k=0}^{L-L_g} [\Delta(\xi, V_k; \xi_m, G_{c,k})\Delta(\xi, V_k'; \xi_n, V_k')] P(\xi_m, t)P(\xi_n, t)$$

Summarizing the construction and the destruction of ξ stated in all four cases, the expected proportion of ξ at time $t + 1$ after the copy-and-paste transposition becomes

$$P(\xi, t+1) = P(\xi, t) - P_A(\xi, t) + P_B(\xi, t) - P_C(\xi, t) + P_D(\xi, t)$$

$$= P(\xi, t) + \frac{p_{copy}}{(L - L_g + 1)^2} \times$$

$$\left\{ -\sum_{c=0}^{L-L_g} \sum_{k=0}^{L-L_g} [1 - \Delta(\xi, V_k; \xi, G_{c,k})] P(\xi, t)P(\xi, t) \right.$$

$$+ \sum_{\xi_n \in S_{\bar{\xi}}} \sum_{c=0}^{L-L_g} \sum_{k=0}^{L-L_g} [\Delta(\xi, V_k; \xi, G_{c,k})\Delta(\xi, V_k'; \xi_n, V_k')] P(\xi, t)P(\xi_n, t) \tag{4.29}$$

$$- \sum_{\xi_m \in S_{\bar{\xi}}} \sum_{c=0}^{L-L_g} \sum_{k=0}^{L-L_g} [1 - \Delta(\xi, V_k; \xi_m, G_{c,k})] P(\xi_m, t)P(\xi, t)$$

$$\left. + \sum_{\xi_m \in S_{\bar{\xi}}} \sum_{\xi_n \in S_{\bar{\xi}}} \sum_{c=0}^{L-L_g} \sum_{k=0}^{L-L_g} [\Delta(\xi, V_k; \xi_m, G_{c,k})\Delta(\xi, V_k'; \xi_n, V_k')] P(\xi_m, t)P(\xi_n, t) \right\}$$

By regrouping the items, the schema evolution equation under the copy-and-paste operation can be derived as

$$P(\xi, t+1) = (1 - p_{copy})P(\xi, t) + \frac{p_{copy}}{(L - L_g + 1)^2}$$

(4.30)

$$\times \sum_{\xi_m \in S_\xi} \sum_{\xi_n \in S_\xi} \sum_{c=0}^{L-L_g} \sum_{k=0}^{L-L_g} \left[\Delta(\xi, V_k; \xi_m, G_{c,k}) \Delta(\xi, V'_k; \xi_n, V'_k) \right] P(\xi_m, t) P(\xi_n, t)$$

4.2.3 Exact Schema Evolution Equation for Cut-and-Paste

Similar to the copy-and-paste operation, the cut-and-paste operation for schemata ξ_m, ξ_n is graphically depicted in Figure 4.3. A transposon of length $1 \le L_g < L$ is cut from ξ_m and pasted onto ξ_n to form a new schema ξ'_n. At the same time, another transposon with the same length is cut from ξ_n and pasted onto ξ_m to form ξ'_m. To insert the transposons into the selected locations (starting with the bit locations k_m and k_n, respectively), some bits in the original schemata have to be shifted accordingly.

Referring to Figure 4.3, different regions are defined as follows:

1. **Transposon region:** G_{c_j,c_i,k_i} indicates the bit locations of the selected transposon cut from ξ_j that fall into the defining length region $[p_1, p_2]$ of ξ after pasting into ξ_i.

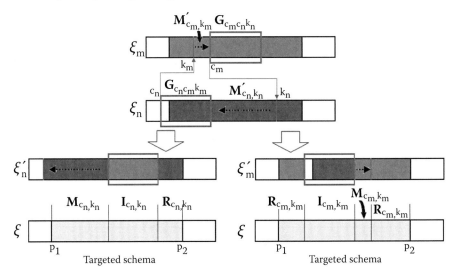

FIGURE 4.3
Cut-and-paste on schemata ξ_m and ξ_n where ξ'_m and ξ'_n are the resultant schemata that will be compared with the targeted schema ξ. (From Tang, K. S., Yin, R. J., Kwong, S., Ng, K. T., Man, K. F., A theoretical development and analysis of jumping gene genetic algorithm, *IEEE Transactions on Industrial Informatics*, 7(3), 2011, 408–418.)

2. **Pasted region**: I_{c_i,k_i} indicates the bit locations within $[p_1, p_2]$ of ξ where the transposon is pasted onto ξ_i.
3. **Moved region**: M_{c_i,k_i} indicates the bit locations within $[p_1, p_2]$ of ξ where some bits of ξ_i are shifted.
4. **Moving region**: M'_{c_i,k_i} indicates the bit locations in ξ_i shifting to M_{c_i,k_i} in ξ'_i.
5. **Unchanged region**: R_{c_i,k_i} indicates the bit locations within $[p_1, p_2]$ where ξ_i remains unaffected by the operation.

The calculations of these regions can be derived as follows:

$$G_{c_j,c_i,k_i} = [c_j + \max(p_1 - L_i, 0), c_j + \min(p_1 + L_d - L_i, L_g - 1)] \tag{4.31}$$

$$I_{c_i,k_i} = [\max(p_1, L_i), \min(p_1 + L_d, L_i + L_g - 1)] \tag{4.32}$$

$$M_{c_i,k_i} = [\max(p_1, k_i + L_g), \min(c_i + L_g - 1, p_1 + L_d)] \\ \cup \; [\max(p_1, c_i), \min(k_i - L_g - 1, p_1 + L_d)] \tag{4.33}$$

$$M'_{c_i,k_i} = [\max(p_1 - L_g, k_i), \min(c_i - 1, p_1 + L_d - L_g)] \\ \cup \; [\max(p_1 + L_g, c_i + L_g), \min(k_i - 1, p_1 + L_d + L_g)] \tag{4.34}$$

$$R_{c_i,k_i} = [p_1, \min(k_i - 1, c_i - 1, p_1 + L_d)] \cup [\max(k_i, c_i + L_g, p_1), p_1 + L_d] \tag{4.35}$$

where $L_i = \mathrm{median}(c_i, k_i - L_g, k_i)$.

Again, according to the choices of ξ_m and ξ_n, three different cases are studied.

Case A

$\xi_m = \xi$ and $\xi_n = \xi$.

To have $\xi'_m \neq \xi$, that is, schema ξ is destroyed, one must have

$$\Delta(\xi, I_{c_m,k_m}; \xi'_m, I_{c_m,k_m})\Delta(\xi, M_{c_m,k_m}; \xi'_m, M_{c_m,k_m}) \neq 1 \\ \Rightarrow \Delta(\xi, I_{c_m,k_m}; \xi, G_{c_n,c_m,k_m})\Delta(\xi, M_{c_m,k_m}; \xi, M'_{c_m,k_m}) \neq 1 \tag{4.36}$$

Similarly, to have $\xi'_n \neq \xi$, one has

$$\Delta(\xi, I_{c_n,k_n}; \xi'_n, I_{c_n,k_n})\Delta(\xi, M_{c_n,k_n}; \xi'_n, M_{c_n,k_n}) \neq 1 \\ \Rightarrow \Delta(\xi, I_{c_n,k_n}; \xi, G_{c_m,c_n,k_n})\Delta(\xi, M_{c_n,k_n}; \xi, M'_{c_n,k_n}) \neq 1 \tag{4.37}$$

By considering all the possible combinations of c_m, c_n and k_m, k_n, the destructive rate PD_A of ξ is then computed as

$$
PD_A = \frac{p_{cut}}{(L-L_g+1)^3} \sum_{c_n=0}^{L-L_g}\sum_{c_m=0}^{L-L_g} \frac{1}{2}\left\{ \sum_{k_m \notin \kappa(c_m)} [1-\Delta(\xi, I_{c_m,k_m}; \xi, G_{c_n,c_m,k_m})\Delta(\xi, M_{c_m,k_m}; \xi, M'_{c_m,k_m})] \right.
$$

$$
\left. + \sum_{k_n \notin \kappa(c_n)} [1-\Delta(\xi, I_{c_n,k_n}; \xi, G_{c_m,c_n,k_n})\Delta(\xi, M_{c_n,k_n}; \xi, M'_{c_n,k_n})] \right\}
$$

$$(4.38)$$

where $\kappa(c_m) = (c_m, c_m + L_g]$, p_{cut} is the operational rate of the cut-and-paste operation, and the factor $\frac{1}{(L-L_g+1)^3}$ indicates the probability of selecting a particular set of (c_m, c_n, k_m) or (c_m, c_n, k_n).

The occurrence probability of case A is given by

$$E_A(t) = P(\xi, t)\, P(\xi, t) \tag{4.39}$$

Hence the expected destruction of ξ in this case is computed as

$$P_A(\xi, t) = PD_A \times E_A(t)$$

$$= \frac{p_{cut}}{(L-L_g+1)^3}$$

$$
\times \sum_{c_n=0}^{L-L_g}\sum_{c_m=0}^{L-L_g} \frac{1}{2}\left\{ \sum_{k_m \notin \kappa(c_m)} [1-\Delta(\xi, I_{c_m,k_m}; \xi, G_{c_n,c_m,k_m})\Delta(\xi, M_{c_m,k_m}; \xi, M'_{c_m,k_m})] \right.
$$

$$
\left. + \sum_{k_n \notin \kappa(c_n)} [1-\Delta(\xi, I_{c_n,k_n}; \xi, G_{c_m,c_n,k_n})\Delta(\xi, M_{c_n,k_n}; \xi, M'_{c_n,k_n})] \right\} P(\xi, t)\, P(\xi, t)
$$

$$(4.40)$$

Case B

$(\xi_m = \xi$ and $\xi_n \in S_{\bar\xi})$ or $(\xi_m \in S_{\bar\xi}$ and $\xi_n = \xi)$.

In this case, one of the schemata is ξ, while another is not. Without duplication, we only consider having $\xi_m = \xi$ and $\xi_n \in S_{\bar\xi}$.

To have ξ destroyed, that is, $\xi'_m \neq \xi$, the condition must be

$$\Delta\left(\xi, I_{c_m,k_m}; \xi'_m, I_{c_m,k_m}\right)\Delta\left(\xi, M_{c_m,k_m}; \xi'_m, M_{c_m,k_m}\right) \neq 1$$

$$\Rightarrow \Delta\left(\xi, I_{c_m,k_m}; \xi_n, G_{c_n,c_m,k_m}\right)\Delta\left(\xi, M_{c_m,k_m}; \xi, M'_{c_m,k_m}\right) \neq 1$$

$$(4.41)$$

On the other hand, ξ is formed if $\xi'_n = \xi$ under the condition specified as follows:

$$\Delta\left(\xi, I_{c_n,k_n}; \xi'_n, I_{c_n,k_n}\right)\Delta\left(\xi, M_{c_n,k_n}; \xi'_n, M_{c_n,k_n}\right)\Delta\left(\xi, R_{c_n,k_n}; \xi'_n, R_{c_n,k_n}\right) \neq 0$$

$$\Rightarrow \Delta\left(\xi, I_{c_n,k_n}; \xi, G_{c_m,c_n,k_n}\right)\Delta\left(\xi, M_{c_n,k_n}; \xi_n, M'_{c_n,k_n}\right)\Delta\left(\xi, R_{c_n,k_n}; \xi_n, R_{c_n,k_n}\right) \equiv \Delta_{IMR} \neq 0$$

$$(4.42)$$

By considering all the possible combinations of c_m, c_n, k_m, k_n, the destructive rate PD_B and the constructive rate PC_C of ξ are derived respectively as

$$PD_B = \frac{p_{cut}}{(L - L_g + 1)^3}$$

$$(4.43)$$

$$\times \sum_{c_n=0}^{L-L_g} \sum_{c_m=0}^{L-L_g} \sum_{k_m \notin \kappa(c_m)} \frac{1}{2}\left[1 - \Delta(\xi, I_{c_m,k_m}; \xi_n, G_{c_n,c_m,k_m})\Delta(\xi, M_{c_m,k_m}; \xi, M'_{c_m,k_m})\right]$$

and

$$PC_B = \frac{p_{cut}}{(L - L_g + 1)^3} \sum_{c_n=0}^{L-L_g} \sum_{c_m=0}^{L-L_g} \sum_{k_n \notin \kappa(c_n)} \frac{\Delta_{IMR}}{2} \qquad (4.44)$$

The occurrence probability of selecting a particular ξ_m and ξ_n, such that $\xi_m = \xi$ and $\xi_n \in S_{\xi}$, is given as

$$E_B(t) = P(\xi, t)P(\xi_n, t) \qquad (4.45)$$

Putting the two situations together, the destruction of ξ in case B is

$$P_{B1}(\xi, t) = 2 \times PD_B \times E_B(t)$$

$$= \frac{p_{cut}}{(L - L_g + 1)^3} \sum_{\xi_n \neq S_{\bar{\xi}}} \sum_{c_n=0}^{L-L_g} \sum_{c_m=0}^{L-L_g} \sum_{k_m \notin \kappa(c_m)} \left[1 - \Delta(\xi, I_{c_m,k_m}; \xi_n, G_{c_n,c_m,k_m})\right. \qquad (4.46)$$

$$\left. \times \Delta(\xi, M_{c_m,k_m}; \xi, M'_{c_m,k_m})\right]P(\xi, t)P(\xi_n, t)$$

and the construction is

$$P_{B2}(\xi, t) = 2 \times PC_B \times E_B(t)$$

$$(4.47)$$

$$= \frac{p_{cut}}{(L - L_g + 1)^3} \sum_{\xi_n \in S_{\bar{\xi}}} \sum_{c_n=0}^{L-L_g} \sum_{c_m=0}^{L-L_g} \sum_{k_n \notin \kappa(c_n)} \Delta_{IMR} P(\xi, t)P(\xi_n, t)$$

Case C

$\xi_m, \xi_n \in S_{\bar{\xi}}$.

The last case is that both schemata are not ξ. The condition to have $\xi'_m = \xi$ is

$$\Delta(\xi, I_{c_m,k_m}; \xi'_m, I_{c_m,k_m}) \Delta(\xi, M_{c_m,k_m}; \xi'_m, M_{c_m,k_m}) \Delta(\xi, R_{c_m,k_m}; \xi'_m, R_{c_m,k_m}) \neq 0$$

$$\Rightarrow \Delta(\xi, I_{c_m,k_m}; \xi_n, G_{c_n,c_m,k_m}) \Delta(\xi, M_{c_m,k_m}; \xi_m, M'_{c_m,k_m}) \Delta(\xi, R_{c_m,k_m}; \xi_m, R_{c_m,k_m}) \equiv \Delta_{IMR_m} \neq 0$$

$$(4.48)$$

Similarly, to have $\xi'_n = \xi$, the condition is

$$\Delta(\xi, I_{c_n,k_n}; \xi'_n, I_{c_n,k_n}) \Delta(\xi, M_{c_n,k_n}; \xi'_n, M_{c_n,k_n}) \Delta(\xi, R_{c_n,k_n}; \xi'_n, R_{c_n,k_n}) \neq 0$$

$$\Rightarrow \Delta(\xi, I_{c_n,k_n}; \xi_m, G_{c_m,c_n,k_n}) \Delta(\xi, M_{c_n,k_n}; \xi_n, M'_{c_n,k_n}) \Delta(\xi, R_{c_n,k_n}; \xi_n, R_{c_n,k_n}) \equiv \Delta_{IMR_n} \neq 0$$

$$(4.49)$$

Therefore, the constructive rate ξ for a particular pair of ξ_m and ξ_n is

$$PC_C = \frac{p_{cut}}{(L-L_g+1)^3} \sum_{c_n=0}^{L-L_g} \sum_{c_m=0}^{L-L_g} \frac{1}{2} \left\{ \sum_{k_m \notin \kappa(c_m)} \Delta_{IMR_m} + \sum_{k_n \notin \kappa(c_n)} \Delta_{IMR_n} \right\}$$

$$(4.50)$$

Together with the occurrence probability of $\xi_m, \xi_n \in S_{\bar{\xi}}$, the construction of ξ in case C can be computed as

$$P_C(\xi,t) = \frac{p_{cut}}{(L-L_g+1)^3} \times \sum_{\xi_m \in S_{\bar{\xi}}} \sum_{\xi_n \in S_{\bar{\xi}}} \sum_{c_n=0}^{L-L_g} \sum_{c_m=0}^{L-L_g}$$

$$(4.51)$$

$$\frac{1}{2} \left\{ \sum_{k_m \notin \kappa(c_m)} \Delta_{IMR_m} + \sum_{k_n \notin \kappa(c_n)} \Delta_{IMR_n} \right\} P(\xi_m,t) P(\xi_n,t)$$

By summing all three cases and regrouping the terms, the schema evolution equation of ξ at time $(t+1)$ under the cut-and-paste operation is

$$P(\xi,t+1) = P(\xi,t) - P_A(\xi,t) - P_{B1}(\xi,t) + P_{B2}(\xi,t) + P_C(\xi,t)$$

$$= (1 - p_{cut}) P(\xi,t)$$

$$(4.52)$$

$$+ \frac{p_{cut}}{(L-L_g+1)^3} \sum_{\xi_m \in S_{\bar{\xi}}} \sum_{\xi_n \in S_{\bar{\xi}}} \sum_{c_m=0}^{L-L_g} \sum_{c_n=0}^{L-L_g} \sum_{k_n \notin \kappa(c_n)} \Delta_{IMR_n} P(\xi_m,t) P(\xi_n,t)$$

where $\kappa(c_n) = (c_n, c_n + L_g)$.

4.3 Theorems of Equilibrium and Dynamical Analysis

Based on the schemata evolution equations of copy-and-paste and cut-and-paste, (4.30) and (4.52), the following two theorems can be derived.

Theorem 4.1: Theorem of Equilibrium for Copy-and-Paste

For any primary schemata competition set S_ξ with order $o(\xi)$, all the schemata in S_ξ will globally asymptotically converge to $1/2^{o(\xi)}$ under a copy-and-paste operation, despite the initial proportion of the schemata in the population. ∎

Theorem 4.2: Theorem of Equilibrium for Cut-and-Paste

For any primary schemata competition set S_ξ with order $o(\xi)$, all the schemata in S_ξ will globally asymptotically converge to $1/2^{o(\xi)}$ under a cut-and-paste operation, despite the initial proportion of the schemata in the population. ∎

The proofs of the theorems are given next.

4.3.1 Distribution Matrix for Copy-and-Paste

Consider the copy-and-paste operation and let $p_{copy} = 1$, for which Equation (4.30) can be rewritten as

$$P(\xi_i, t+1) = \sum_{\xi_m \in S_\xi} \sum_{\xi_n \in S_\xi} a_{mn}^{(i)} \times P(\xi_m, t) \times P(\xi_n, t) \text{ for } \xi_i \in S_\xi, \ i =, 1, 2, \ldots M, \quad (4.53)$$

where

$$a_{mn}^{(i)} = K \sum_{c=0}^{L-L_g} \sum_{k=0}^{L-L_g} [\Delta(\xi_i, V_k; \xi_m, G_{c,k}) \Delta(\xi_i, V_k'; \xi_n, V_k')] \quad (4.54)$$

with

$$K = \frac{1}{(L - L_g + 1)^2}.$$

Definition 4.6

The *distribution matrix* for schema $\xi_i \in S_\xi$ under the copy-and-paste operation is defined as

$$
A^{(i)} = \begin{bmatrix}
a_{11}^{(i)} & a_{12}^{(i)} & \cdots & a_{1M}^{(i)} \\
a_{21}^{(i)} & a_{22}^{(i)} & \cdots & a_{2M}^{(i)} \\
\vdots & & \ddots & \\
a_{M1}^{(i)} & a_{M2}^{(i)} & \cdots & a_{MM}^{(i)}
\end{bmatrix}
\tag{4.55}
$$

where M is the size of S_ξ.

Remark 4.5

For each schema $\xi_i \in S_\xi$, $A^{(i)}$ only depends on the JG length L_g. ■

Remark 4.6

Based on the definition given in Equation (4.54), $a_{mn}^{(i)} \in [0,1]$.
   We define a state vector at time t as

$$
Y(t) = \begin{bmatrix}
P(\xi_1, t) \\
P(\xi_2, t) \\
\vdots \\
P(\xi_M, t)
\end{bmatrix}
\tag{4.56}
$$

then (4.53) can be rewritten as

$$
P(\xi_i, t+1) = Y'(t)A^{(i)}Y(t) = y_i(t+1) \text{ for } i = 1, 2, \ldots, M,
\tag{4.57}
$$

where $Y'(t)$ denotes the transpose of $Y(t)$, and $y_i(t)$ is the ith element of $Y(t)$. Based on the definition of $P(\xi_i, t)$, one has

$$
\sum_{i=1}^{M} y_i(t) = \sum_{i=1}^{M} P(\xi_i, t) = 1 \quad \forall t
\tag{4.58}
$$

Equation (4.57) is regarded as the state transition equation, revealing the dynamics of the copy-and-paste transposition. ■

Example 4.3

Consider $S_\xi = \{*00***, *01***, *10***, *11***\}$ and $L_g = 2$; based on Definition 4.6, we can obtain the following distribution matrices:

$$A^{(1)} = \begin{bmatrix} 0.78 & 0.24 & 0.24 & 0.1 \\ 0.64 & 0.14 & 0.14 & 0.04 \\ 0.64 & 0.14 & 0.14 & 0.04 \\ 0.54 & 0.08 & 0.08 & 0.02 \end{bmatrix}, \quad A^{(2)} = \begin{bmatrix} 0.1 & 0.64 & 0.04 & 0.18 \\ 0.16 & 0.66 & 0.06 & 0.16 \\ 0.16 & 0.66 & 0.06 & 0.16 \\ 0.18 & 0.64 & 0.04 & 0.1 \end{bmatrix},$$

$$A^{(3)} = \begin{bmatrix} 0.1 & 0.04 & 0.64 & 0.18 \\ 0.16 & 0.06 & 0.66 & 0.16 \\ 0.16 & 0.06 & 0.66 & 0.16 \\ 0.18 & 0.04 & 0.64 & 0.1 \end{bmatrix}, \quad A^{(4)} = \begin{bmatrix} 0.02 & 0.08 & 0.08 & 0.54 \\ 0.04 & 0.14 & 0.14 & 0.64 \\ 0.04 & 0.14 & 0.14 & 0.64 \\ 0.1 & 0.24 & 0.24 & 0.78 \end{bmatrix}.$$

Let the state at time t be

$$Y(t) = \begin{bmatrix} P(\xi_1, t) \\ P(\xi_2, t) \\ P(\xi_3, t) \\ P(\xi_4, t) \end{bmatrix} = \begin{bmatrix} 0.1 \\ 0.2 \\ 0.3 \\ 0.4 \end{bmatrix},$$

Using Equation (4.57), one can then obtain

$$P(\xi_1, t+1) = Y'(t)A^{(1)}Y(t) = 0.1396,$$

$$P(\xi_2, t+1) = Y'(t)A^{(2)}Y(t) = 0.2164,$$

$$P(\xi_3, t+1) = Y'(t)A^{(3)}Y(t) = 0.2764,$$

$$P(\xi_4, t+1) = Y'(t)A^{(4)}Y(t) = 0.3676.$$

Therefore, the state at time $t + 1$ is

$$Y(t+1) = \begin{bmatrix} 0.1396 \\ 0.2164 \\ 0.2764 \\ 0.3676 \end{bmatrix}.$$

4.3.2 Distribution Matrix for Cut-and-Paste

Similarly, let us consider the cut-and-paste operation. Assuming that $p_{cut} = 1$, Equation (4.52) becomes

$$P(\xi_i, t+1) = \sum_{\xi_m \in S_\xi} \sum_{\xi_n \in S_\xi} b_{mn}^{(i)} P(\xi_m, t) P(\xi_n, t) \text{ for } \xi_i \in S_\xi, i = 1, 2, \cdots, M \quad (4.59)$$

where

$$b_{mn}^{(i)} = \frac{1}{(L - L_g + 1)^3} \sum_{c_n=0}^{L-L_g} \sum_{c_m=0}^{L-L_g} \sum_{k_n \notin \kappa(c_n)} \Delta_{IMR_n} \quad (4.60)$$

and the distribution matrix can be defined as follows:

$$B^{(i)} = \begin{bmatrix} b_{11}^{(i)} & b_{12}^{(i)} & \cdots & b_{1M}^{(i)} \\ b_{21}^{(i)} & b_{22}^{(i)} & \cdots & b_{2M}^{(i)} \\ & & \ddots & \\ b_{M1}^{(i)} & b_{M2}^{(i)} & \cdots & b_{MM}^{(i)} \end{bmatrix} \quad (4.61)$$

such that

$$y_i(t+1) = Y'(t) B^{(i)} Y(t) \quad \text{for } i = 1, 2, \ldots, M \quad (4.62)$$

where $Y(t)$ is defined as given in Equation (4.57).

4.3.3 Lemmas

Before presenting the proof of Theorems 4.1 and 4.2, some lemmas are given. The proofs of these lemmas are given in Appendix A.

Lemma 4.1

Considering a quadruple $q = (\xi_m, \xi_n, c, k)$, a set of q is defined as

$$Q = \{q : c \in [0, L - L_g]; k \in [0, L - L_g]; \xi_m \in S_\xi; \xi_n \in S_\xi\},$$

and a function is specified by

$$f^{(i)}(q) = \delta(f_T(\xi_i, V_k), f_T(\xi_m, G_{c,k})) \times \delta(f_T(\xi_i, V_k'), f_T(\xi_n, V_k')),$$

there exists a bijective mapping $g: Q \to Q$ such that $\tilde{q} = g(q)$ and $f^{(j)}(\tilde{q}) = f^{(i)}(q)$. ∎

Lemma 4.2

$$\sum_{m=1}^{M}\sum_{n=1}^{M} a_{mn}^{(i)} = M, \quad \forall i \in \{1,2,\cdots,M\}.$$

■

Lemma 4.3

$$\sum_{i=1}^{M} a_{mn}^{(i)} = 1, \quad \forall m,n \in \{1,2,\cdots,M\}.$$

■

Lemma 4.4

$$\sum_{\xi_n \in Z} a_{mn}^{(i)} - \sum_{\xi_n \in Z'} a_{mn}^{(i)} = \pm\frac{N_0}{L-L_g-1} \equiv C_{row}$$

where Z is Z_{odd} or Z_{even}, and N_0 is the number of possible pasting positions in ξ_i such that the pasted region contains don't-care bits (*) only. ■

Lemma 4.5

$$\sum_{\xi_m \in Z} a_{mn}^{(i)} - \sum_{\xi_m \in Z'} a_{mn}^{(i)} = \pm\frac{N_{all}}{(L-L_g-1)^2} \equiv C_{col}$$

where Z is Z_{odd} or Z_{even}, and N_{all} is the number of possible pasting positions in ξ_i such that the pasted region includes all the actual bits. ■

Lemma 4.6

$$-1 < C_{row} + C_{col} < 1.$$

■

Lemma 4.7

Consider a distribution matrix $A^{(i)} = \{a_{mn}^{(i)}\}$ with size $M \times M$; one has

$$Q_1 = \sum_{\xi_m \in Z}\sum_{\xi_n \in Z} a_{mn}^{(i)} = \frac{M}{4}(1 + C_{row} + C_{col});$$

$$Q_2 = \sum_{\xi_m \in Z'}\sum_{\xi_n \in Z'} a_{mn}^{(i)} = \frac{M}{4}(1 - C_{row} - C_{col});$$

$$Q_3 = \sum_{\xi_m \in Z} \sum_{\xi_n \in Z'} a_{mn}^{(i)} = \frac{M}{4}(1 - C_{row} + C_{col});$$

$$Q_4 = \sum_{\xi_m \in Z'} \sum_{\xi_n \in Z} a_{mn}^{(i)} = \frac{M}{4}(1 + C_{row} - C_{col})$$

where Z is Z_{odd} or Z_{even}. ∎

Lemma 4.8

$$\sum_{m=1}^{M} \sum_{n=1}^{M} b_{mn}^{(i)} = M, \quad \forall i \in \{1,2,\cdots,M\}.$$

∎

Lemma 4.9

$$\sum_{i=1}^{M} b_{mn}^{(i)} = 1, \quad \forall m,n \in \{1,2,\cdots,M\}.$$

∎

Lemma 4.10

$$\sum_{\xi_n \in Z} b_{mn}^{(i)} - \sum_{\xi_n \in Z'} b_{mn}^{(i)} = \pm \frac{N_0'}{(L - L_g + 1)^2} \equiv C_{row}',$$

where Z is Z_{odd} or Z_{even}, and N_0' is the number of possible combinations of cutting and pasting positions such that the transposon and moving regions in ξ_i contain don't-care bits (*) only. ∎

Lemma 4.11

$$\sum_{\xi_m \in Z} b_{mn}^{(i)} - \sum_{\xi_m \in Z'} b_{mn}^{(i)} = \pm \frac{N_{all}}{(L - L_g + 1)^2} \equiv C_{col}'$$

where Z is Z_{odd} or Z_{even}, and N_{all} is the number of possible cutting positions such that the transposon region in ξ_i includes all the actual bits. ∎

Lemma 4.12

$$-1 < C'_{row} + C'_{col} < 1.$$ ∎

Lemma 4.13

Considering a distribution matrix $B^{(i)} = \{b^{(i)}_{mn}\}$ with size $M \times M$, one has

$$Q_1 = \sum_{\xi_m \in Z} \sum_{\xi_n \in Z} b^{(i)}_{mn} = \frac{M}{4}(1 + C'_{row} + C'_{col});$$

$$Q_2 = \sum_{\xi_m \in Z'} \sum_{\xi_n \in Z'} b^{(i)}_{mn} = \frac{M}{4}(1 - C'_{row} - C'_{col});$$

$$Q_3 = \sum_{\xi_m \in Z} \sum_{\xi_n \in Z'} b^{(i)}_{mn} = \frac{M}{4}(1 - C'_{row} + C'_{col});$$

$$Q_4 = \sum_{\xi_m \in Z'} \sum_{\xi_n \in Z} b^{(i)}_{mn} = \frac{M}{4}(1 + C'_{row} - C'_{col})$$

where Z is Z_{odd} or Z_{even}. ∎

4.3.4 Proof of Theorem 4.1

First, by mathematical induction, the case when $M = 2$ is considered. Recalling Equation (4.57),

$$P(\xi_i, t+1) = Y'(t)A^{(i)}Y(t) \tag{4.63}$$

where

$$Y(t) = \begin{bmatrix} P(\xi_1, t) \\ P(\xi_2, t) \end{bmatrix}$$

and

$$A^{(i)} = \begin{bmatrix} a^{(i)}_{11} & a^{(i)}_{12} \\ a^{(i)}_{21} & a^{(i)}_{22} \end{bmatrix}.$$

Based on the definition of Q_i in Lemma 4.7, one can have $a_{11}^{(i)} = Q_1$, $a_{12}^{(i)} = Q_3$, $a_{21}^{(i)} = Q_4$, and $a_{22}^{(i)} = Q_2$. Hence,

$$P(\xi_1, t+1) = Q_1 P^2(\xi_1, t) + (Q_3 + Q_4)P(\xi_1, t)P(\xi_2, t) + Q_2 P^2(\xi_2, t)$$

$$= Q_1 P^2(\xi_1, t) + (Q_3 + Q_4)P(\xi_1, t)(1 - P(\xi_1, t)) + Q_2(1 - P(\xi_1, t))^2$$

$$= (Q_1 + Q_2 - Q_3 - Q_4)P^2(\xi_1, t) + (Q_3 + Q_4 - 2Q_2)P(\xi_1, t) + Q_2 \quad (4.64)$$

$$= (C_{row} + C_{col})P(\xi_1, t) + \frac{1}{2}(1 - C_{row} - C_{col})$$

By iterating backward, one has

$$P(\xi_1, t) = (C_{row} + C_{col})^t P(\xi_1, 0) + \frac{1}{2}(1 - C_{row} - C_{col}) \times \frac{1 - (C_{row} + C_{col})^t}{(1 - (C_{row} + C_{col})}$$

$$= \frac{1}{2} + (C_{row} + C_{col})^t \left(P(\xi_1, 0) - \frac{1}{2} \right) \quad (4.65)$$

Based on Lemma 4.6, one has $(C_{row} + C_{col})^t \to 0$ when $t \to \infty$. From Equation (4.65), $P(\xi_1, \infty) = \frac{1}{2}$ and $P(\xi_2, \infty) = 1 - P(\xi_1, \infty) = \frac{1}{2}$. Therefore, the theorem is true for all S_ξ with $o(\xi) = 1$.

Now, it is assumed that the theorem is true for any S_ξ with $o(\xi) = n$ and $M = 2^n$, that is, $Y = \frac{1}{M} 1_{M \times 1}$ or $P(\xi_i) = \frac{1}{M} \ \forall \xi_i \in S_\xi$, where $P(\xi_i)$ and Y represent $P(\xi_i, \infty)$ and $Y(\infty)$, respectively, for simplicity.

Consider any primary schemata competition set S_φ where $o(\varphi_i) = n+1$. It is easily proved that there exists $n + 1$ subsets of S_ξ with $o(\xi) = n$ and $M = 2^n$ such that the locations of the actual bits in S_ξ are also the locations of actual bits in S_φ. Mathematically, $f_L(\xi) \subset f_L(\varphi)$.

S_φ can be divided into two subgroups $S_{\varphi,0}$ and $S_{\varphi,1}$, which are formed by the following rules:

The actual bits of the first element in $S_{\varphi,0}$ are all zeros, while the other elements are found by flipping the last actual bit and changing the other bits according to Gray coding.

The actual bits of the first element in $S_{\varphi,1}$ are all zeros except the last actual bit, while the other elements are found by flipping the last actual bit and changing the other bits according to Gray coding.

Remark 4.7

Obviously, it can be proved that $S_{\varphi,0} \cup S_{\varphi,1} = S_\varphi$ and $S_{\varphi,0} \cap S_{\varphi,1} = \emptyset$. Moreover, $size(S_{\varphi,0}) = size(S_{\varphi,1}) = M$. ∎

Remark 4.8

The elements in $S_{\varphi,i}$ are cyclic, and two consecutive elements have two bits difference. In addition, any two elements in $S_{\varphi,i}$ will also have at least two bits of difference (otherwise, the rule of Gray coding will be violated). ∎

Remark 4.9

Based on Remark 4.8, by changing any particular actual bit location of $S_{\varphi,0}$ (or $S_{\varphi,0}$) into the wild character '*', a subgroup of schemata S_ξ is formed. (It is easily proved as every element will then have different actual bits according to Remark 4.8.) Therefore, $P(\varphi_i)$ must be the same for all $\varphi_i \in S_{\varphi,0}$ (or $\varphi_i \in S_{\varphi,1}$). ∎

Example 4.4

Let $S_\varphi = \{*0**00*0, *0**00*1, \cdots, *1**11*1\}$; we can divide it into two groups:

$$
S_{\varphi,0} = \left\{
\begin{array}{l}
*0**00*0, \\
*0**01*1, \\
*0**11*0, \\
*0**10*1, \\
*1**10*0, \\
*1**11*1, \\
*1**01*0, \\
*1**00*1
\end{array}
\right\}
\quad
S_{\varphi,1} = \left\{
\begin{array}{l}
*0**00*1, \\
*0**01*0, \\
*0**11*1, \\
*0**10*0, \\
*1**10*1, \\
*1**11*0, \\
*1**01*1, \\
*1**00*0
\end{array}
\right\}.
$$

Remarks 4.7–4.9 can be easily verified. Furthermore, all elements in $S_{\varphi,0}$ belong to Z_{even}, whereas all elements in $S_{\varphi,1}$ belong to Z_{odd} in this example.

Without loss of generality, let us consider the case of $o(\varphi) = n + 1$, which is even. Hence, if $\varphi_i \in S_{\varphi,0}$, we have $\varphi_i \in Z_{even}$. Similarly, if $\varphi_i \in S_{\varphi,1}$, we have $\varphi_i \in Z_{odd}$.

Based on Remark 4.9, one has

$$P(\varphi_i) = C_0 \text{ for } \varphi_i \in S_{\varphi,0} \tag{4.66}$$

$$P(\varphi_j) = C_1 \text{ for } \varphi_j \in S_{\varphi,1} \tag{4.67}$$

where C_0 and C_1 are constants. Using Equation (4.53) and assuming that $\varphi_i \in S_{\varphi,0}$, we have

$$P(\varphi_i) = \sum_{m=1}^{M'} \sum_{n=1}^{M'} a_{mn}^{(i)} P(\varphi_m) P(\varphi_n)$$

$$
\begin{aligned}
C_0 &= \sum_{\varphi_m \in Z_{even}} \sum_{\varphi_n \in Z_{even}} a_{mn}^{(i)} P(\varphi_m) P(\varphi_n) + \sum_{\varphi_m \in Z_{even}} \sum_{\varphi_n \in Z_{odd}} a_{mn}^{(i)} P(\varphi_m) P(\varphi_n) \\[2mm]
&\quad + \sum_{\varphi_m \in Z_{odd}} \sum_{\varphi_n \in Z_{even}} a_{mn}^{(i)} P(\varphi_m) P(\varphi_n) + \sum_{\varphi_m \in Z_{odd}} \sum_{\varphi_n \in Z_{odd}} a_{mn}^{(i)} P(\varphi_m) P(\varphi_n) \\[2mm]
&= \sum_{\varphi_m \in Z_{even}} \sum_{\varphi_n \in Z_{even}} a_{mn}^{(i)} C_0^2 + \sum_{\varphi_m \in Z_{even}} \sum_{\varphi_n \in Z_{odd}} a_{mn}^{(i)} C_0 C_1 \\[2mm]
&\quad + \sum_{\varphi_m \in Z_{odd}} \sum_{\varphi_n \in Z_{even}} a_{mn}^{(i)} C_1 C_0 + \sum_{\varphi_m \in Z_{odd}} \sum_{\varphi_n \in Z_{odd}} a_{mn}^{(i)} C_1^2 \\[2mm]
&= Q_1 C_0^2 + (Q_3 + Q_4) C_0 C_1 + Q_2 P^2 C_1^2 \\[2mm]
&= Q_1 C_0^2 + (Q_3 + Q_4) C_0 \left(\frac{1}{M} - C_0 \right) + Q_2 \left(\frac{1}{M} - C_0 \right)^2 \\[2mm]
&= (Q_1 + Q_2 - Q_3 - Q_4) C_0^2 + (Q_3 + Q_4 - 2Q_2) \frac{C_0}{M} + \frac{1}{M^2} Q_2 \\[2mm]
&= (C_{row} + C_{col}) C_0 + \frac{1}{2M} (1 - C_{row} - C_{col}) \\[2mm]
&= \frac{1}{2M}
\end{aligned}
$$

(4.68)

Since $\sum\limits_{i=1}^{M'} P(\varphi_i) = 1$, $C_1 = \frac{1}{M}(1 - C_0 \times M) = \frac{1}{2M} = C_0$. Therefore, $P(\varphi_i) = \frac{1}{2M}, \forall \varphi_i \in S_\varphi$. By mathematical induction, the proof is completed.

4.3.5 Proof of Theorem 4.2

By using Lemmas 4.8 to 4.13, the proof of Theorem 4.2 can be obtained by following the same steps given in the proof of Theorem 4.1.

4.4 Simulation Results and Analysis

In previous sections, the convergence to equilibrium of a primary schemata competition set based on copy-and-paste and cut-and-paste operations was mathematically proved. For illustration, some simulations follow.

4.4.1 Simulation 4.1: Existence of Equilibrium

Consider primary schemata competition set $S_\xi = \{* * * \#^* \#^* * * \#^* * * * * *\}$, where # is the actual bit.

Figures 4.4a and 4.4b depict the proportions of schemata in the population against generations under copy-and-paste and cut-and-paste operations

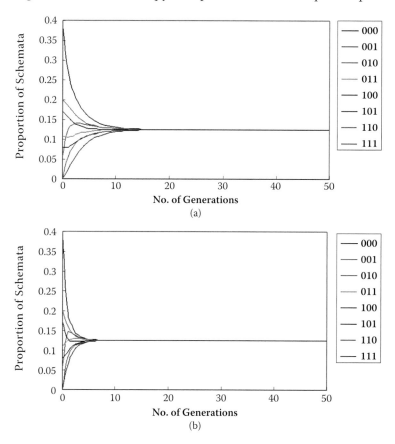

(a)

(b)

FIGURE 4.4 (SEE COLOR INSERT.)
The proportion of schemata against generations using (a) copy-and-paste and (b) cut-and-paste operations, where 000 stands for ***0*0***0****** and so on. (From Yin, J. J., Yeung, S. H., Tang, W. K. S., Man, K. S., Kwong, S., Enhancement of multiobjective search: A Jumping-genes approach, in *Proceedings of IEEE International Symposium on Industrial Electronics*, Vigo, Spain, June 2007, 1855–1858.)

based on Equations (4.29) and (4.51), respectively. The initial proportion of each schema is randomly assigned, and it is clearly observed that the proportion goes to the equilibrium of 0.125 ($1/2^3$), despite the initial states. It is true even though some schemata in set S_ξ may not exist initially.

4.4.2 Simulation 4.2: Primary Schemata Competition Sets with Different Orders

In Simulation 4.2, the growth of different order primary schemata competition sets in the population is recorded. For the order 1, 2, 3, 4 competition sets, the representatives are taken as ***#************, ***#*****#******, ***#*#***#******, and ***#*#**##******, respectively, where # is the actual bit. The initial proportion of each schema of order 4 is randomly assigned, while the initial portions of the other three sets can be calculated. For example, if the proportions of schemata ***0*0**00****** and ***0*0**10****** are 0.1 and 0.05, respectively, the proportion of schema ***0*0***0****** is then determined by $0.1 + 0.05 = 0.15$.

Figures 4.5a and 4.5b show that, using either the copy-and-paste or the cut-and-paste operation, the primary schemata competition sets of different orders ultimately reach their own equilibria. It is also noticed that it takes longer for a higher-order primary schemata competition set to reach equilibrium than for its corresponding lower-order set.

4.5 Discussion

4.5.1 Assumptions

It should be remarked that the proof of the theorem of equilibrium is based on the assumption that the don't-care bit (*) has an equal chance to be 0 or 1. This assumption is not common in other schema theories. A don't-care bit is not to be compared with any definite bits in traditional genetic operations; therefore, its chance to be 0 or 1 is not important. However, this assumption is required in our proofs; fortunately, if the initial population is large enough and contains sufficient randomness, it is more or less true.

4.5.2 Implications

As demonstrated by many previous reports, techniques such as fitness sharing [4,5,11] and niche crowding [2,18] adopted in an MOEA are essential to maintain the diversity in the population and prevent a crowded population. From this point of view, there are at least two implications for the theorems of equilibrium. First, any possible individual would have the same occurrence

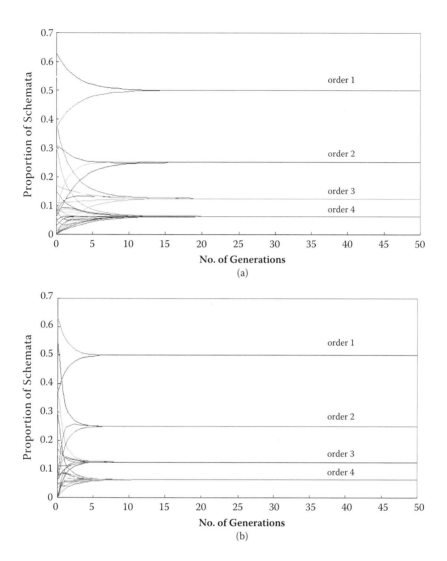

FIGURE 4.5

The proportion of schemata of different orders against generations using (a) copy-and-paste and (b) cut-and-paste operations. (From Tang, K. S., Yin, R. J., Kwong, S., Ng, K. T., Man, K. F., A theoretical development and analysis of jumping gene genetic algorithm, *IEEE Transactions on Industrial Informatics*, 7(3), 2011, 408–418.)

frequency; therefore, the global searching ability of these two JG operations is justified. It should be emphasized that the capability of searching globally by these two JG operations is rigorously proven by mathematics instead of relying on a statistical behavior of randomness such as for the mutation in conventional GA. Second, the JG operations tend to respect the survival rates of all the schemata equally despite the current distribution of the schemata. Hence, the supergood chromosomes that dominate the population will be suppressed and premature convergence can be prevented. More discussion about the theorems of equilibrium and their effect is presented in further sections.

4.5.3 Destruction and Construction

From the derivation of the schema evolution equations, we can observe that both the destruction and the construction of schemata play important roles. With a higher destruction rate and a lower construction rate, the number of chromosomes belonging to such a schema will decrease sharply. In contrast, a schema with a low destructive rate and high constructive rate will dominate the whole population.

Let us take copy-and-paste as an example. According to Equations (4.14) and (4.22), the destructive rate of ξ_i under a copy-and-paste operation is given by

$$PD = \frac{p_{copy}}{(L-L_g+1)^2} \sum_{c=0}^{L-L_g} \sum_{k=0}^{L-L_g} \left[1 - \Delta(\xi_i, V_k; \xi_m, G_{c,k})\right] \tag{4.69}$$

where $\xi_m \in S_{\xi_i}$. Similarly, based on Equations (4.18) and (4.26), the constructive rate can be formulated as

$$PC = \frac{p_{copy}}{(L-L_g+1)^2} \sum_{c=0}^{L-L_g} \sum_{k=0}^{L-L_g} \left[\Delta(\xi_i, V_k; \xi_m, G_{c,k})\Delta(\xi_i, V_k'; \xi_n, V_k')\right] \tag{4.70}$$

where $\xi_m \in S_{\xi_i}, \xi_n \in S_{\bar{\xi}_i}$.

Recall the definition of $a_{mn}^{(i)}$ in Equation (4.54); the destructive rate Equation (4.69) becomes

$$PD = \frac{p_{copy}}{(L-L_g+1)^2} \sum_{c=0}^{L-L_g} \sum_{k=0}^{L-L_g} [1 - \Delta(\xi_i, V_k; \xi_m, G_{c,k})]$$

$$= p_{copy}\left(1 - \frac{1}{(L-L_g+1)^2} \sum_{c=0}^{L-L_g} \sum_{k=0}^{L-L_g} \Delta(\xi_i, V_k; \xi_m, G_{c,k})\right) \tag{4.71}$$

$$= p_{copy}\left(1 - a_{mi}^{(i)}\right)$$

and the constructive rate Equation (4.70) becomes

$$PC = \frac{p_{copy}}{(L - L_g + 1)^2} \sum_{c=0}^{L-L_g} \sum_{k=0}^{L-L_g} \left[\Delta(\xi_i, V_k; \xi_m, G_{c,k}) \Delta(\xi_i, V'_k; \xi_n, V'_k) \right]$$

(4.72)

$$= p_{copy} \left(a_{mn}^{(i)} \right)$$

As shown in Equations (4.71) and (4.72), the destructive and constructive rates depend on $a_{mn}^{(i)}$, which, based on its definition and Remark 4.5, depends on the JG length L_g, for which in general a short JG length is more preferable.

4.5.4 Finite Population Effect

The theorem of equilibrium is based on the assumption that the population size is large enough. However, in real situations the population size is limited or even small. To illustrate the effect of a finite population size, we conducted the following simulations:

Consider a 16-bit chromosome and the primary schemata competition set $S_\xi = \{***\#* \#***\#******\}$. A population of 100 chromosomes is randomly generated, and the population undergoes the two JG operations only (no selection, crossover, or mutation). Figures 4.6a and 4.6b depict the dynamics of primary schemata competition set S_ξ in this finite pool under copy-and-paste and cut-and-paste operations, respectively. As the population size is now limited, each schema will compete with other schemata in the same competition set to reach equilibrium. As a result, the proportion of each schema will fluctuate around the equilibrium, and the state of equilibrium cannot be maintained. Obviously, the deviation from equilibrium for cut-and-paste is smaller than that for copy-and-paste.

Although the state of equilibrium cannot be maintained in a finite population, all schemata have chances to be constructed during JG operations. Therefore, JG operations possess the property of ergodicity. For illustration, another example was carried out by considering a population of 100 chromosomes of length 16. Figure 4.7 depicts the result of searching abilities of various operations, including copy-and-paste, cut-and-paste, uniform crossover, two-point crossover, mutations with an operational rate of 0.5 and 0.05, and random generation.

The initial population was obtained in two different ways. In Figure 4.7a, it was randomly generated. Throughout the generations, the population in the next generation was solely generated by the designated operations, and the occurrence of each possible individual was recorded. As shown in Figure 4.7a, the cut-and-paste operation generates all possible individuals (a total of 2^{16}) at least once in the fewest generations, while the number of generations used for the copy-and-paste operation was comparable with that

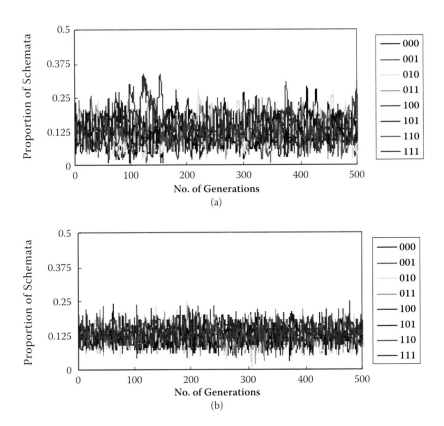

FIGURE 4.6 (SEE COLOR INSERT.)
The proportion of schemata against generations in a finite population using (a) copy-and-paste and (b) cut-and-paste operations. (From Tang, K. S., Yin, R. J., Kwong, S., Ng, K. T., Man, K. F., A theoretical development and analysis of jumping gene genetic algorithm, *IEEE Transactions on Industrial Informatics*, 7(3), 2011, 408–418.)

for mutation with a rate of 0.5. It is obvious that crossover operations were incapable of achieving the task, while random generation also failed as it was only statistically possible.

A similar result was obtained for the second case, as shown in Figure 4.7b, where all the chromosomes in the population were identical but with randomly generated genes. It is remarked that no new individual can be obtained by crossover operations as the parents are the same.

4.5.5 The Effect of the JG in a GA

As indicated by the theorem of equilibrium, JG operations try to maintain all the schemata in equal proportion despite the current distribution of schemata. This effect can be further demonstrated by introducing the cut-and-paste or copy-and-paste operation into a conventional GA.

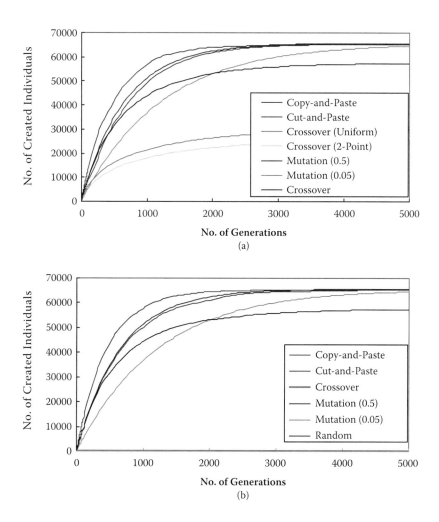

FIGURE 4.7 (SEE COLOR INSERT.)
The searching abilities of different operations in a finite population with initial population (a) randomly generated and (b) of identical chromosomes. (From Tang, K. S., Yin, R. J., Kwong, S., Ng, K. T., Man, K. F., A theoretical development and analysis of jumping gene genetic algorithm, *IEEE Transactions on Industrial Informatics*, 7(3), 2011, 408–418.)

Consider a particular primary schemata competition set {000; 001; 010; 011; 100; 101; 110; 111} with the corresponding average fitness {0; 4; 2; 6; 1; 5; 3; 7}, respectively. Figure 4.8a shows the proportion of schemata against generations based on the conventional GA. It is observed that the best schema will dominate the population, while the rest survive mainly due to mutation operations. This pattern of dynamics highly depends on the presence of the fitness function and the selection process.

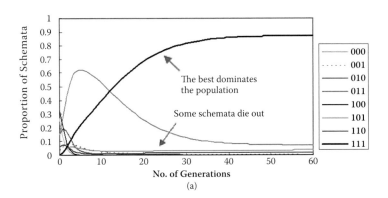

(a)

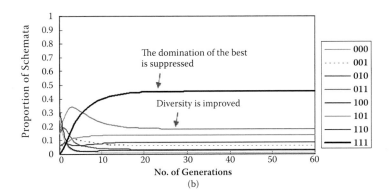

(b)

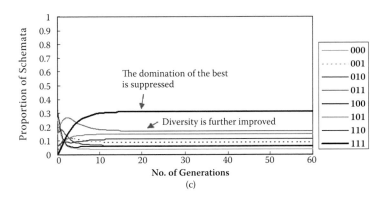

(c)

FIGURE 4.8

The proportion of schemata in the population against generations using a (a) GA, (b) GA with copy and paste, and (c) GA with cut-and-paste. The operational rates of crossover, mutation, and JG operations are 1.0, 0.01, and 0.1, respectively. (From Yin, J. J., Yeung, S. H., Tang, W. K. S., Man, K. S., Kwong, S., Enhancement of multiobjective search: A Jumping-genes approach, in *Proceedings of IEEE International Symposium on Industrial Electronics*, Vigo, Spain, June 2007, 1855–1858.)

Figures 4.8b and 4.8c show the cases when the copy-and-paste and cut-and-paste operations were incorporated into the conventional GA, respectively. It was clearly demonstrated that the occurrence of the best schema was suppressed, and diversity was maintained. The effect of suppression was more significant with the use of the cut-and-paste operation as compared with the result using the copy-and-paste operation.

References

1. Altenberg, L., The schema theorem and Price's theorem, in *Proceedings of Third Workshop on Foundations of Genetic Algorithms*, Estes Park, CO, July 31–August 2, 1994, 23–49.
2. Cedeno, W., Vemuri, V. R., Analysis of speciation and niching in the multi-niche crowding GA, *Theoretical Computer Science*, 229(1–2), 177–197, 1999.
3. Das, R., Whitley, D., The only challenging problems are deceptive: Global search by solving order-1 hyperplanes, in *Proceedings of International Conference on Genetic Algorithms*, San Diego, CA, July 1991, 166–173.
4. Della Cioppa, A., De Stefano, C., Marcelli, A., On the role of population size and niche radius in fitness sharing, *IEEE Transactions on Evolutionary Computation*, 8(6), 580–592, 2004.
5. Della Cioppa, A., De Stefano, C., Marcelli, A., Where are the niches? Dynamic fitness sharing, *IEEE Transactions on Evolutionary Computation*, 11(4), 453–465, 2007.
6. Fogel, D. B., Ghozeil, A., Schema processing under proportional selection in the presence of random effects, *IEEE Transactions on Evolutionary Computation*, 1(4), 290–293, 1997.
7. Fogel, D. B., Ghozeil, A., The schema theorem and the misallocation of trials in the presence of stochastic effects, in *Proceedings of the Seventh Annual Conference on Evolutionary Programming*, San Diego, CA, March 1998, 313–321.
8. Goldberg, D. E., *Genetic Algorithms in Search, Optimization, and Machine Learning*, Reading, MA: Addison-Wesley, 1989.
9. Holland, J. H., *Adaptation in Natural and Artificial Systems*, Ann Arbor: University of Michigan Press, 1975.
10. Li, M. Q., Kou, J. S., The schema deceptiveness and deceptive problems of genetic algorithms, *Science in China—Series F*, 44(5), 342–350, 2001.
11. Sareni, B., Krahenbuhl, L., Fitness sharing and niching methods revisited, *IEEE Transactions on Evolutionary Computation*, 2(3), 97–106, 1998.
12. Spears, W. M., *Evolutionary Algorithms: The Role of Mutation and Recombination*, New York: Springer, 2000.
13. Stephens, C. R., Waelbroeck, H., Effective degrees of freedom in genetic algorithms and the block hypothesis, in *Proceedings of Seventh International Conference on Genetic Algorithms*, East Lansing, MI, July 1997, 34–40.
14. Stephens, C. R., Waelbroeck, H., Aguirre, R., Schemata as building blocks: Does size matter? in *Proceedings of Fifth Workshop on Foundations of Genetic Algorithms*, Amsterdam, September 1998, 117–133.

15. Stephens, C. R., Waelbroeck, H., Schemata evolution and building blocks, *Evolutionary Computation*, 7(2), 109–124, 1999.
16. Tang, K. S., Kwong, S., Man, K. F., A jumping genes paradigm: Theory, verification and applications, *IEEE Circuits and Systems Magazine*, 8(4), 18–36, 2008.
17. Tang, K. S., Yin, R. J., Kwong, S., Ng, K. T., Man, K. F., A theoretical development and analysis of jumping gene genetic algorithm, *IEEE Transactions on Industrial Informatics*, 7(3), 2011, 408–418.
18. Vemuri, V. R., Cedeno, W., A new genetic algorithm for multi-objective optimization in water resource management, in *Proceedings of IEEE International Conference on Evolutionary Computation*, Perth, Australia, November 1995, 1:495–500.
19. Whitley, D., Fundamental principles of deception in genetic search." In *Proceedings of 1st Workshop on Foundations of Genetic Algorithms*, Bloomington Campus, IN, July 1990, 221–241.
20. Yin, R. J., Schema theorem for computational gene transposition and performance analysis, Ph.D. thesis, City University of Hong Kong, 2010.
21. Yin, J. J., Yeung, S. H., Tang, W. K. S., Man, K. S., Kwong, S., Enhancement of multiobjective search: A Jumping-genes approach, in *Proceedings of IEEE International Symposium on Industrial Electronics*, Vigo, Spain, June 2007, 1855–1858.

5

Performance Measures on Jumping Gene

5.1 Convergence Metric: Generational Distance

The generational distance convergence metric α measures the degree of closeness between the sets of the nondominated solutions and the true Pareto-optimal solutions. Its mathematical representation [18] is given by

$$\alpha = \frac{1}{N} \left(\sum_{i=1}^{N} d_i^p \right)^{\frac{1}{p}} \tag{5.1}$$

where d_i is the distance between the ith nondominated solution and its nearest true Pareto-optimal solution, and N is the total number of solutions in the nondominated set.

It was recommended in Zitzler [20] that a value of $p = 1$ is more appropriate due to its simplicity and averaging characteristics [7]. Then, Equation (5.1) becomes

$$\alpha = \frac{1}{N} \sum_{i=1}^{N} d_i \tag{5.2}$$

The steps for calculating α are as follows:

1. Find a set of, say, 500 uniformly spaced nondominated solutions lying on the true Pareto-optimal front.
2. For each nondominated solution I_i, find the Euclidean distances d_{iq} between the solution i and each true Pareto-optimal solution P_q in the space of the objective values using the following formula:

$$d_{iq} = \sqrt{\sum_{m=1}^{M} \left(f_m(I_i) - f_m(P_q) \right)^2} \tag{5.3}$$

where M is the total number of objective functions, and $f_m(I_i)$ and $f_m(P_q)$ are the values of the mth objective function of the nondominated solution I_i and a true Pareto-optimal solution P_q, respectively.

3. Set $d_i = \min(d_{iq})$.

4. Compute the metric α by Equation (5.2).

For the optimal case, the metric value is zero when all the nondominated solutions are exactly equal to the selected true Pareto-optimal solutions. Also, a smaller metric value indicates that the nondominated front is closer to the true Pareto-optimal front, that is, there is better convergence.

5.2 Convergence Metric: Deb and Jain Convergence Metric

For most real-world applications, the huge search space results in impracticality for finding true Pareto-optimal solutions. Therefore, the use of the generational distance convergence metric described in Section 5.1 may not be practicable. Alternatively, a set of reference solutions is used to form the pseudo-Pareto-optimal front. This has been employed in another metric called the Deb and Jain convergence metric [8] so that the convergence metric can be computed.

Assuming that a group of algorithms is considered, the procedures for producing the reference solutions are as follows: (1) generate, say, 50 sets of nondominated solutions from each algorithm and (2) rank all nondominated solutions obtained by these algorithms. The final single set of nondominated solutions is then regarded as the reference solution set.

The mathematical representation of this metric (denoted β) is the same as that of the generational distance. For $p = 1$, one has

$$\beta = \frac{1}{N}\sum_{i=1}^{N}d_i \tag{5.4}$$

where d_i is the normalized distance between the ith nondominated solution and its nearest reference solution, and N is the total number of solutions in the nondominated set.

The steps to calculate β are as follows:

1. For each nondominated solution I_i, calculate the normalized Euclidean distances d_{ir} between the solution I_i and each reference solution R_r by

$$d_{ir} = \sqrt{\sum_{m=1}^{M}\left(\frac{f_m(I_i)-f_m(R_r)}{f_m(R_{max})-f_m(R_{min})}\right)^2} \tag{5.5}$$

where M is the total number of objective functions; $f_m(I_i)$ and $f_m(R_r)$ return the mth objective values of the nondominated solution I_i and a reference solution R_r, respectively; and $f_m(R_{max})$ and $f_m(R_{min})$ are the maximum and minimum values of the mth objective function within all solutions in the reference set, respectively.

2. Set $d_i = \min(d_{ir})$.

3. Compute the metric β by (5.4).

Similar to the generational distance, a smaller value of β reveals that the non-dominated front is closer to the reference front, that is, there is better convergence.

5.3 Diversity Metric: Spread

The spread diversity metric evaluates the distribution of a nondominated solution set spreading along the true Pareto-optimal or the reference front [7], which is calculated as

$$\gamma = \frac{\sum\limits_{m=1}^{M} d_m^e + \sum\limits_{i=1}^{N-1} \left| d_i - \bar{d} \right|}{\sum\limits_{m=1}^{M} d_m^e + (N-1) \cdot \bar{d}} \tag{5.6}$$

where M is the total number of objective functions; d_m^e is the distance between an extreme solution in the true Pareto-optimal set and that in the nondominated set for the mth objective function; d_i is the distance between the ith and $(i + 1)$th nondominated solutions (i.e., neighbor nondominated solutions); $\bar{d}$ is the average of all d_i; and N is the total number of solutions in the nondominated set.

The procedures for calculating γ are as follows:

1. Calculate all Euclidean distances d_i for $i = 1, 2, \cdots, (N-1)$:

$$d_i = \sqrt{\sum\limits_{m=1}^{M} \left(f_m(I_i) - f_m(I_{i+1}) \right)^2} \tag{5.7}$$

where M is the total number of objective functions, and $f_m(I_i)$ and $f_m(I_{i+1})$ return the mth objective values of the nondominated solution I_i and the nondominated solution I_{i+1}, respectively.

2. Calculate the average value of d_i:

$$\bar{d} = \frac{1}{(N-1)} \sum_{i=1}^{N-1} d_i \qquad (5.8)$$

3. Obtain the value of d_m^e for $m = 1, 2, \cdots, M$:

$$d_m^e = \left| f_m(I_{ext}) - f_m(P_{ext}) \right| \qquad (5.9)$$

where $f_m(I_{ext})$ and $f_m(P_{ext})$ are the values of the mth objective values of extreme solutions in the nondominated set and in the true Pareto-optimal set, respectively.

4. Compute the metric γ by Equation (5.6).

If the nondominated solutions found are uniformly distributed [i.e., $d_i = \bar{d}$ for $i = 1, 2, \cdots, (N-1)$] and the nondominated extreme solutions are equal to the true Pareto-optimal extreme solutions (i.e., $\sum_{m=1}^{M} d_m^e = 0$), the value of γ will be zero. This is the most ideal case for the distribution of the nondominated solutions.

Regarding the distribution in which $d_i = \bar{d} \ \forall i$, Equation (5.6) becomes

$$\gamma = \frac{\sum\limits_{m=1}^{M} d_m^e}{\sum\limits_{m=1}^{M} d_m^e + (N-1) \cdot \bar{d}} \qquad (5.10)$$

If the nondominated solutions are crowded in a small region, $\sum_{m=1}^{M} d_m^e \neq 0$, one has $0 < \gamma < 1$.

In addition, $\gamma \geq 1$ represents an even-worse distribution. A smaller value of γ means that the nondominated front is more widely spread and more uniformly distributed along the true Pareto-optimal or reference front (i.e., there is better diversity).

5.4 Diversity Metric: Extreme Nondominated Solution Generation

The "knee" of the nondominated front and its extremes should be equally important. However, the extreme solutions are preferable in some applications, and their existence is critical for multiobjective optimization. For example, suppose that there are two contradicting objectives, cost and performance; yet investing higher cost can acquire better performance and vice versa. When a system requires at least 90% performance and minimum cost simultaneously, it is most likely that only extreme non-dominated solutions are considered.

A new unary performance metric called extreme nondominated solution generation is proposed as another way for measuring diversity. This metric computes the total number of extreme nondominated solutions found after a simulation run. Its calculation procedures are discussed next.

Consider the coordinates of two extreme solutions e_1 and e_2 of the true Pareto-optimal or reference front in the hyperspace of objective values (with dimension M) as $(f_1(e_1), f_2(e_1), \cdots, f_M(e_1))$ and $(f_1(e_2), f_2(e_2), \cdots, f_M(e_2))$, respectively; the radii r_1 and r_2 of two spheres with centers at e_1 and e_2, respectively, are computed as:

$$r_1 = \sqrt{\sum_{m=1}^{M} \left(\mu_1 f_m(e_1)\right)^2} \tag{5.11}$$

and

$$r_2 = \sqrt{\sum_{m=1}^{M} \left(\mu_2 f_m(e_2)\right)^2} \tag{5.12}$$

where μ_i is a scaling factor to control the size of the circles.

To count the total number of extreme nondominated solutions falling inside the two hyperspheres for a number of nondominated sets, the following steps can be used:

1. Set $C = 0$, where C is the total number of extreme nondominated solutions falling inside the two hyperspheres.
2. Set $j = 1$ (j is the counter of the nondominated set).
3. Set $i = 1$ (i is the counter of the nondominated solution for the jth nondominated set).
4. For the ith nondominated solution with objective values of $(f_1(I_i), f_2(I_i), \ldots, f_M(I_i))$ in the jth nondominated set, calculate the distances between the solution i and the extreme solution e_1 and e_2 by

$$d_1 = \sqrt{\sum_{m=1}^{M} \left(f_m(I_i) - f_m(e_1)\right)^2} \tag{5.13}$$

and

$$d_2 = \sqrt{\sum_{m=1}^{M} \left(f_m(I_i) - f_m(e_2) \right)^2} \tag{5.14}$$

5. If either of the following two conditions is satisfied, (a) $d_1 \le r_1$, (b) $d_2 \le r_2$, $C = C + 1$.

6. Increment i by 1, and if $i \le N$ where N is the total number of solutions in the jth nondominated set, go to step 4.

7. Increment j by 1, and if $j \le S$ where S is the total number of nondominated sets, go to step 3.

8. Output the value of C.

A larger metric value indicates that the nondominated set offers more extreme nondominated solutions to decision makers. Note that this metric cannot be used alone to assess the diversity of nondominated sets. This is because a large metric value does not guarantee that the obtained nondominated solutions are uniformly distributed. Thus, it must be adopted together with other diversity metrics at the same time.

5.5 Binary ε-Indicator

It was criticized that the unary metrics described in Sections 5.1–5.4 are incapable of indicating whether one nondominated set is better than another [22,23]. Therefore, a binary ε-indicator has been proposed [23] and used to identify the better performer when a pair of nondominated solution sets is examined.

Consider a minimization problem with M positive objectives; an objective vector $f^1 = (f^1_1, f^1_2, \ldots, f^1_M)$ is said to ε-dominate another objective vector $f^2 = (f^2_1, f^2_2, \ldots, f^2_M)$, written as $f^1 \succeq_\varepsilon f^2$, if and only if

$$\forall 1 \le i \le M : f^1_i \le \varepsilon \cdot f^2_i \tag{5.15}$$

for a given $\varepsilon > 0$.

The binary ε-indicator I_ε is defined as $I_\varepsilon(A,B) = \inf_{\varepsilon \in \Re} \{ \forall f^2 \in B, \exists f^1 \in A : f^1 \succeq_\varepsilon f^2 \}$ for any two nondominated solution sets A and B. It can be calculated as

$$\varepsilon_{f^1,f^2} = \max_{1 \le i \le M} \frac{f^1_i}{f^2_i} \forall f^1 \in A, f^2 \in B \tag{5.16}$$

$$\varepsilon_{f^2} = \min_{f^1 \in A} \varepsilon_{f^1, f^2} \; \forall \; f^2 \in B \tag{5.17}$$

$$I_\varepsilon(A, B) = \max_{f^2 \in B} \varepsilon_{f^2} \tag{5.18}$$

or equivalently

$$I_\varepsilon(A, B) = \max_{f^2 \in B} \min_{f^1 \in A} \max_{1 \le i \le M} \frac{f_i^1}{f_i^2} \tag{5.19}$$

While comparing two solution sets A and B, three different cases can result:

Case I: $I_\varepsilon(A, B) \le 1$ and $I_\varepsilon(B, A) > 1 \Rightarrow$ Set A is better than Set B.

Case II: $I_\varepsilon(A, B) > 1$ and $I_\varepsilon(B, A) \le 1 \Rightarrow$ Set B is better than Set A.

Case III: $I_\varepsilon(A, B) > 1$ and $I_\varepsilon(B, A) > 1 \Rightarrow$ Sets A and B are incomparable.

This gives rise to three scenarios when two multiobjective optimization algorithms, say MO_A and MO_B, are compared:

1. Favorable to the MO_A: $\Gamma_{N_I} > \Gamma_{N_{II}}$ and $\Gamma_{N_I} > \Gamma_{N_{III}}$
2. Unfavorable to the MO_A: $\Gamma_{N_{II}} > \Gamma_{N_I}$ and $\Gamma_{N_{II}} > \Gamma_{N_{III}}$
3. Inconclusive: $\Gamma_{N_{III}} > \Gamma_{N_I}$ and $\Gamma_{N_{III}} > \Gamma_{N_{II}}$

where Γ_{N_I}, $\Gamma_{N_{II}}$, and $\Gamma_{N_{III}}$ are the number of occurrences of Cases I, II, and III, with the solution sets A and B obtained by MO_A and MO_B, respectively.

5.6 Statistical Test Using Performance Metrics

Judging the convergence and diversity performance of a multiobjective evolutionary algorithm (MOEA) through performance metrics is an important issue in multiobjective optimization [6,7]. Since MOEAs are probabilistic search techniques, their offered solution quality is not guaranteed to be good for all simulation runs.

It turns out that if only one or several sets of nondominated solutions produced by a particular MOEA are examined, a conclusion about the performance of the MOEA is probably wrong. The reason is that the acquired results can be biased. Hence, a statistical test is necessary to assess the performance of MOEAs [6]. To carry out the statistical test, a sufficiently large number of simulation runs, say $K = 50$ runs, is needed. Given that a nondominated solution set is acquired from each run, the value of a particular unary performance metric can then be calculated. Last, the mean and standard deviation of K different metric values can be obtained as the basic performance indicator.

Furthermore, the statistical comparison of pairs of nondominated fronts is adopted for the binary ε-indicator. Each MOEA should perform K simulation runs to generate K nondominated solution sets. To obtain this metric, each solution in the K nondominated sets of an MOEA is compared with those of another MOEA, one by one. Therefore, there are a total of K^2 comparisons; each comparison results in one of the three comparison cases mentioned in Section 5.5. Finally, the number of occurrences for each case is counted and compared.

5.7 Jumping Gene Verification and Results

Just like other genetic operations, such as crossover and mutation, jumping gene (JG) transposition can easily be integrated into any general framework of an MOEA (see Section 2.2 in Chapter 2 for details of an MOEA). Based on our studies, note that the inclusion of JG transposition in the nondominated sorting genetic algorithm 2 (NSGA2) [9] can result in a good search performance. For differentiation, this algorithm is referred to as JG in the following discussion.

To verify the effectiveness of JG in multiobjective optimization, the most straightforward approach is to use some well-known benchmark test functions for evaluating its performance. The true Pareto-optimal fronts of these test functions should be obtainable with various characteristics (e.g., concave, convex, disconnected, etc.). In the following, eight unconstrained (SCH [14]; FON [10]; POL [13]; ZIT1, -2, -3, -4, and -6 [21]) and five constrained (DEB [9], BEL [1], SRIN [15], TAN [16], and BINH [2]) benchmark test functions were chosen. The descriptions of these benchmark test functions are given in Appendix B.

As mentioned in Chapter 3, JG transposition is workable for different data types. Thus, we adopted binary chromosome representation for the 13 test functions, while real-number chromosome representations were also applied for the test functions ZIT1, ZIT2, ZIT3, ZIT4, and ZIT6. Further details on the encoding and decoding techniques of binary and real-number chromosomes are provided in Appendix C.

5.7.1 JG Parameter Study

The success of the JG relies on three important parameters: jumping rate, number of transposons, and length of transposon. Any variation of these parameters can affect its performance on convergence and diversity.

An investigation was performed for verifying the effect of each parameter; two unconstrained test functions (FON and POL) and two constrained test functions (SRIN and TAN) are used here for demonstration.

Since a four-dimensional plot with axes (jumping rate, number of transposons, length of transposon, and convergence/diversity metric value) is impossible, the number of transposons N and the length of transposon L as the two related indications of JGs were combined to form a single parameter called the parameter index. This allows us to present the results in three-dimensional (3D) plots. The correspondence between the parameter index and the associated number of transposons and length of transposons is given in Tables 5.1 and 5.2.

The 3D plots for average values of convergence and diversity metrics against different parameter indexes and jumping rates for different test functions are shown in Figure 5.1. The average convergence or diversity metric value of each bar in the figure was obtained by averaging the 50 convergence or diversity metric values acquired in 50 simulation runs, respectively.

These results revealed that lower jumping rates (0.005–0.1) enhanced the convergence performance; the effect was not significant for the number and length of transposons. However, good performance could be obtained with

TABLE 5.1

Correspondence between the Parameter Index and the Associated Total Number and Length of Transposons of the Binary-Coded JG for Test Function FON

Parameter Index	N	L	Parameter Index	N	L
1	1	1	19	4	4
2	1	2	20	4	6
3	1	4	21	4	8
4	1	6	22	4	10
5	1	8	23	6	1
6	1	10	24	6	2
7	1	20	25	6	4
8	1	40	26	6	6
9	1	60	27	8	1
10	2	1	28	8	2
11	2	2	29	8	4
12	2	4	30	10	1
13	2	6	31	10	2
14	2	8	32	10	4
15	2	10	33	20	1
16	2	20	34	20	2
17	4	1	35	40	1
18	4	2	36	60	1

TABLE 5.2

Correspondence between the Parameter
Index and the Associated Total Number and
Length of Transposons of the Binary-Coded
JG for Test Functions POL, SRIN, and TAN

Parameter Index	N	L	Parameter Index	N	L
1	1	1	15	4	1
2	1	2	16	4	2
3	1	4	17	4	4
4	1	6	18	4	6
5	1	8	19	6	1
6	1	10	20	6	2
7	1	20	21	6	4
8	1	40	22	8	1
9	2	1	23	8	2
10	2	2	24	10	1
11	2	4	25	10	2
12	2	6	26	20	1
13	2	8	27	40	1
14	2	10			

both the number and length of transposons less than 10. This verifies the manifestation of low transposition rates at which the host remains unharmed [11]. However, when both the jumping rate and number and length of transposons were increased to a certain higher value, poor convergence was evident.

On the other hand, variations of the jumping rate and the number and length of transposons did not affect the diversity performance greatly. In conclusion, value selection of the three parameters based on the mentioned ranges is recommended for JG users to acquire good performance.

5.7.2 Comparisons with Other MOEAs

This section compares the JG with other commonly used MOEAs described in Chapter 2. To perform statistical comparisons, 50 simulation runs were performed for each MOEA in each test function. Table 5.3 shows the parameters of MOEAs utilized in the simulations.

In addition, the number of fitness function evaluations (denoted by $E = P \times G$, where P is the population size and G is the maximum number of generations or iterations) was varied for different test functions. However, for a particular test function, the number of fitness function evaluations E was fixed for all algorithms to allow fair comparisons. For example, referring to Table 5.3, the number of fitness function evaluations for the unconstrained test function SCH was $E = 100 \times 20 = 2000$.

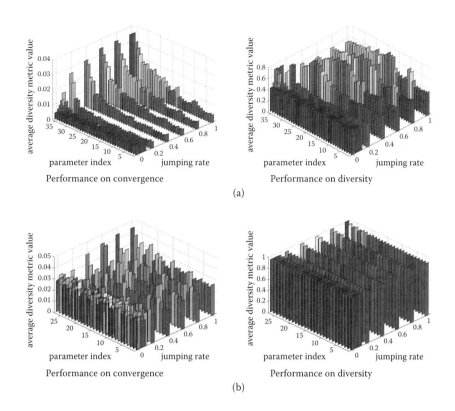

FIGURE 5.1 (SEE COLOR INSERT.)
3D plots for average values of convergence and diversity metrics against different parameter indexes and jumping rates for different test functions. (a) Unconstrained test function: FON; (b) unconstrained test function: POL; (c) constrained test function: SRIN; (d) constrained test function: TAN. (From Chan, T. M., Man, K. F., Kwong, S., Tang, K. S., A jumping gene paradigm for evolutionary multiobjective optimization, *IEEE Transactions on Evolutionary Computation*, 12(2), 143–159, 2008.)

The results for different algorithms under different performance metrics are presented in the following discussion.

5.7.2.1 Mean and Standard Deviation of Generational Distance for Evaluating Convergence

The means and standard deviations of the generational distance convergence metric for the unconstrained test functions appear in Table 5.4. From the table, the JG found the most convergent nondominated solutions for the test functions ZIT3 and ZIT4 (in binary coding) and ZIT1–ZIT4 and ZIT6 (in real-number coding). Also, the JG was the first runner-up for the remaining test functions.

The means and standard deviations of the generational distance for the constrained test functions are given in Table 5.5. From the table, although the JG

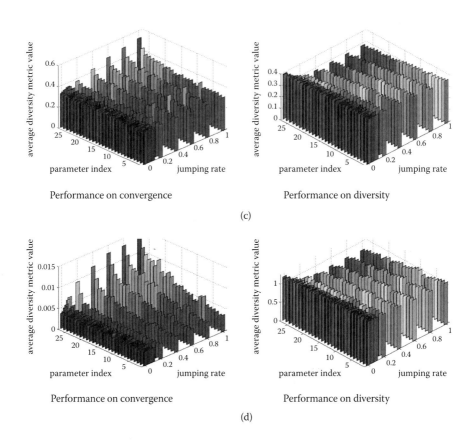

Performance on convergence   Performance on diversity

(c)

Performance on convergence   Performance on diversity

(d)

FIGURE 5.1 (CONTINUED)

did not attain the best convergence for any test function, the difference between its score and that of the winner for each test function was small. Therefore, it can be concluded that the JG showed good performance for convergence.

5.7.2.2 Mean and Standard Deviation of Spread for Evaluating Diversity

Tables 5.6 and 5.7 list the means and standard deviations of the diversity metric spread for the unconstrained test functions with binary and real-number coding, respectively. As can be observed, the JG was capable of finding the most diverse nondominated solutions for the test functions SCH, FON, POL, ZIT4 (in binary coding) and ZIT1, ZIT2, and ZIT4 (in real-number coding). For the remaining test functions, the difference between its score and that of the winner was small.

The means and standard deviations of the spread for the constrained test functions are tabulated in Table 5.8. The JG found the most

TABLE 5.3

Parameters of MOEAs

Parameter	Value/Type
Chromosome encoding method	Binary and real representation
Population size (MOGA, NPGA2, NSGA2, SPEA2, JG)	100
Population size (PAES)	1
Population size (MICROGA)	4
Maximum generations (MOGA, NPGA2, NSGA2, SPEA2, JG)	SCH, FON, POL, DEB, BEL, SRIN, TAN, BINH: 20 ZIT1[a], ZIT2[a], ZIT3[a]: 80 ZIT4[a]: 200 ZIT6[a]: 150
Maximum iterations (PAES)	SCH, FON, POL, DEB, BEL, SRIN, TAN, BINH: 2,000 ZIT1[a], ZIT2[a], ZIT3[a]: 8,000 ZIT4[a]: 20,000 ZIT6[a]: 15,000
Maximum iterations (MICROGA)	SCH, FON, POL, DEB, BEL, SRIN, TAN, BINH: 500 ZIT1[a], ZIT2[a], ZIT3[a]: 2,000 ZIT4[a]: 5,000 ZIT6[a]: 3,750
Crossover type	Uniform crossover
Crossover rate	0.8
Mutation rate	$1/N$ or $1/L$ where N is the total number of variables for real-coded MOEAs and L is the total number of bits in a chromosome for binary-coded MOEAs
Jumping rate (JG)	0.04
Number of transposons (JG)	1
Length of transposons (JG)	3
Archive size (SPEA2, PAES)	100
Depth (PAES)	4
Size of external memory (MICROGA)	100
Size of population memory (MICROGA)	80
Percentage of non-replaceable memory (MICROGA)	0.25
Replacement cycle (MICROGA)	Every 25 iterations
Number of subdivisions of the adaptive grid (MICROGA)	25
Number of iterations to achieve nominal convergence (MICROGA)	4

Source: Data from Chan, T. M., Man, K. F., Kwong, S., Tang, K. S., A jumping gene paradigm for evolutionary multiobjective optimization, *IEEE Transactions on Evolutionary Computation*, 12(2), 143–159, 2008.

Note: NPGA2, niched Pareto genetic algorithm 2; PAES, Pareto archived evolution strategy; SPEA2, strength Pareto evolutionary algorithm 2.

[a] Same maximum generations (iterations) for binary-coded and real-coded MOEAs.

TABLE 5.4

Means and Standard Deviations of the Generational Distance for Unconstrained Test Functions

Test Function	MOGA	NPGA2	NSGA2	SPEA2	PAES	MICROGA	JG
SCH	1.63178	0.00788	0.01083	0.02072	**0.00208**	1110.11	0.00392
	(6.99304)	(0.01784)	(0.01892)	(0.07066)	**(0.00351)**	(4252.05)	(0.00445)
FON	0.01812	**0.01290**	0.01473	0.01367	0.08741	0.05874	0.01349
	(0.00851)	(0.00570)	(0.00495)	(0.00536)	(0.07542)	(0.02101)	(0.00598)
POL	0.05812	0.06243	0.05011	**0.04299**	0.99976	0.21435	0.04361
	(0.06925)	(0.07250)	(0.04996)	**(0.03404)**	(1.64908)	(0.14311)	(0.04206)
ZIT1	0.04444	0.03574	0.03412	0.04036	**0.00494**	2.31672	0.03386
(binary)	(0.00892)	(0.00867)	(0.00827)	(0.00828)	(0.02132)	(0.17970)	**(0.00697)**
ZIT2	0.07996	0.05591	0.05365	0.06032	**0.00162**	3.26253	0.05171
(binary)	(0.01845)	(0.01262)	(0.01189)	(0.01154)	**(0.00189)**	(0.19828)	(0.01171)
ZIT3	0.04627	0.02725	0.02876	0.03534	0.04480	2.18547	**0.02647**
(binary)	(0.01451)	(0.00972)	(0.01100)	(0.01016)	(0.05578)	(0.14426)	**(0.00884)**
ZIT4	6.63284	7.60903	7.15519	7.45696	13.3502	81.4667	**5.83309**
(binary)	(2.98656)	(3.81820)	(3.15991)	(3.49560)	(5.04150)	(10.3024)	**(2.53999)**
ZIT6	0.35795	0.14714	0.15218	0.19644	**0.04066**	6.36652	0.09530
(binary)	**(0.05484)**	(0.09597)	(0.09511)	(0.07452)	(0.07390)	(0.17106)	(0.09862)
ZIT1	0.12911	0.10382	0.11010	0.11222	0.06335	2.17251	**0.05004**
(real)	(0.01555)	(0.01583)	(0.01537)	(0.01476)	(0.02125)	(0.13842)	**(0.01400)**
ZIT2	0.24297	0.20004	0.19478	0.20013	0.10562	3.04937	**0.08177**
(real)	(0.03670)	(0.02543)	(0.03402)	(0.02392)	**(0.02043)**	(0.17496)	(0.02368)
ZIT3	0.06636	0.04612	0.04781	0.05197	0.04332	2.01183	**0.01851**

(real)	(0.01469)	(0.01318)	(0.01183)	(0.01211)	(0.02120)	(0.14565)	**(0.00706)**
ZIT4	0.55196	0.46020	0.30273	0.28212	0.63499	71.86735	**0.012794**
(real)	(0.32172)	(0.39848)	(0.13446)	(0.15333)	(0.57676)	(8.63943)	**(0.02087)**
ZIT6	1.93393	1.93976	1.85939	1.95742	1.77444	6.05983	**1.06337**
(real)	(0.16917)	(0.15781)	**(0.12349)**	(0.17091)	(0.20176)	(0.21104)	(0.33144)

Source: Data from Chan, T. M., Man, K. F., Kwong, S., Tang, K. S., A jumping gene paradigm for evolutionary multiobjective optimization, *IEEE Transactions on Evolutionary Computation*, 12(2), 143–159, 2008.

Note: The best result for test function is marked in bold, and the values in parentheses represent the standard deviation.

TABLE 5.5

Means and Standard Deviations of the Generational Distance for Constrained Test Functions

Test Function	MOGA	NPGA2	NSGA2	SPEA2	PAES	MICROGA	JG
DEB	0.00787	**0.00676**	0.01081	0.01014	0.01824	0.01842	0.01008
	(0.00229)	(0.00171)	(0.00262)	(0.00365)	(0.02965)	(0.00358)	**(0.00170)**
BEL	0.01348	0.01301	0.01240	0.01445	**0.00902**	0.09445	0.01279
	(0.00396)	(0.00261)	(0.00194)	(0.00205)	(0.00417)	(0.02769)	**(0.00152)**
SRIN	**0.33370**	0.34227	0.40167	0.41859	17.3122	1.82348	0.41630
	(0.07496)	(0.08955)	**(0.06420)**	(0.07971)	(24.2239)	(0.72811)	(0.07284)
TAN	0.01352	**0.00958**	0.01189	0.01044	0.05525	0.02776	0.01149
	(0.00385)	**(0.00237)**	(0.00630)	(0.00335)	(0.03728)	(0.00744)	(0.00521)
BINH	0.31397	0.30102	0.34141	**0.28119**	19.7117	1.31411	0.31083
	(0.12715)	(0.13577)	(0.24490)	**(0.11206)**	(23.7084)	(0.78961)	(0.13814)

Source: Data from Chan, T. M., Man, K. F., Kwong, S., Tang, K. S., A jumping gene paradigm for evolutionary multiobjective optimization, *IEEE Transactions on Evolutionary Computation*, 12(2), 143–159, 2008.

Note: The best result for each test function is marked in bold, and the values in parentheses represent the standard deviation.

TABLE 5.6

Means and Standard Deviations of the Spread for Unconstrained Test Functions with Binary Coding

Test Function	MOGA	NPGA2	NSGA2	SPEA2	PAES	MICROGA	JG
SCH	1.02523	1.09615	0.79028	0.84460	1.24120	0.94102	**0.76548**
	(0.14631)	**(0.12280)**	(0.27601)	(0.21778)	(0.28551)	(0.14680)	(0.28694)
FON	1.03046	1.06488	0.81581	0.89288	1.18890	0.84067	**0.79632**
	(0.09428)	(0.08976)	(0.10672)	(0.07915)	**(0.06091)**	(0.17055)	(0.12059)
POL	1.24790	1.30040	0.98009	1.06714	1.45101	1.16311	**0.97765**
	(0.08416)	(0.09810)	**(0.03127)**	(0.03670)	(0.17554)	(0.08594)	(0.03300)
ZIT1	0.63691	0.61230	**0.41442**	0.52405	1.26137	0.84746	0.41465
(binary)	(0.10156)	(0.08824)	(0.05896)	(0.09220)	(0.06556)	**(0.04018)**	(0.05094)
ZIT2	0.61292	0.58163	**0.47624**	0.54716	1.26247	0.89727	0.50918
(binary)	(0.11556)	(0.08569)	(0.07003)	(0.08752)	(0.08099)	**(0.05266)**	(0.12495)
ZIT3	0.86572	0.83810	**0.82077**	0.83004	1.11195	0.84011	0.82535
(binary)	(0.08209)	(0.05907)	(0.09101)	(0.07053)	(0.23000)	**(0.03607)**	(0.06827)
ZIT4	0.95457	0.98082	0.88591	0.90604	1.11646	0.88144	**0.87744**
(binary)	**(0.02219)**	(0.06604)	(0.03558)	(0.04790)	(0.08184)	(0.08978)	(0.04379)
ZIT6	**0.79677**	0.96855	0.82626	0.89355	1.38634	0.93809	0.88508
(binary)	(0.04076)	(0.15051)	(0.16014)	(0.14645)	(0.14814)	**(0.03384)**	(0.21032)

Source: Data from Chan, T. M., Man, K. F., Kwong, S., Tang, K. S., A jumping gene paradigm for evolutionary multiobjective optimization, *IEEE Transactions on Evolutionary Computation*, 12(2), 143–159, 2008.

Note: The best result for each test function is marked in bold, and the values in parentheses represent the standard deviation.

TABLE 5.7

Means and Standard Deviations of the Spread for Unconstrained Test Functions with Real Number Coding

Test Function	MOGA	NPGA2	NSGA2	SPEA2	PAES	MICROGA	JG
ZIT1	0.56998	0.56315	0.54173	0.56256	0.71198	0.83987	**0.51694**
(real)	(0.05726)	(0.04825)	(0.05107)	(0.04346)	(0.15148)	**(0.04098)**	(0.06099)
ZIT2	0.69375	0.65685	0.65005	0.66054	0.70422	0.90731	**0.57713**
(real)	(0.05570)	(0.05319)	(0.06307)	(0.04572)	(0.10218)	**(0.03819)**	(0.06253)
ZIT3	0.66761	0.67783	**0.66309**	0.68165	0.70329	0.82258	0.66627
(real)	(0.07318)	(0.05143)	(0.06146)	(0.06935)	(0.08939)	**(0.04561)**	(0.07855)
ZIT4	0.90134	0.83851	0.70980	0.81572	1.16767	0.87446	**0.36373**
(real)	(0.13221)	(0.10636)	(0.14454)	(0.14291)	(0.27960)	(0.09259)	**(0.06138)**
ZIT6	0.99920	0.99063	1.00089	1.03269	1.13532	**0.92468**	0.95879
(real)	(0.04478)	(0.03754)	(0.03752)	(0.06267)	(0.08332)	**(0.02896)**	(0.08229)

Source: Data from Chan, T. M. Man, K. F., Kwong, S., Tang, K. S., A jumping gene paradigm for evolutionary multiobjective optimization, *IEEE Transactions on Evolutionary Computation*, 12(2), 143–159, 2008.

Note: The best result for each test function is marked in bold, and the values in parentheses represent the standard deviation.

TABLE 5.8

Means and Standard Deviations of the Spread for Constrained Test Functions

Test Function	MOGA	NPGA2	NSGA2	SPEA2	PAES	MICROGA	JG
DEB	1.23439	1.2824	**1.00997**	1.10916	1.34785	1.01390	1.02604
	(0.11148)	(0.10698)	**(0.07196)**	(0.08286)	(0.20127)	(0.09970)	(0.09947)
BEL	0.78913	0.78612	0.38570	0.48759	1.51643	0.64472	**0.37570**
	(0.05983)	(0.06286)	(0.02923)	(0.03960)	(0.07172)	(0.06951)	**(0.02594)**
SRIN	0.93024	0.95896	0.39539	0.50727	1.37574	0.97593	**0.38802**
	(0.08007)	(0.08668)	(0.03431)	**(0.03244)**	(0.15613)	(0.09951)	(0.03557)
TAN	1.03719	1.11753	1.13388	1.11497	**0.79309**	0.83511	1.13543
	(0.11038)	(0.10733)	**(0.09879)**	(0.11324)	(0.24614)	(0.10710)	(0.12711)
BINH	1.04469	1.07929	**0.55134**	0.69212	1.26679	0.98301	0.55704
	(0.12719)	(0.12185)	(0.11730)	**(0.09136)**	(0.18122)	(0.17276)	(0.14308)

Source: Data from Chan, T. M., Man, K. F., Kwong, S., Tang, K. S., A jumping gene paradigm for evolutionary multiobjective optimization, *IEEE Transactions on Evolutionary Computation*, 12(2), 143–159, 2008.

Note: The best result for each test function is marked in bold, and the values in parentheses represent the standard deviation.

diverse nondominated solutions for the test functions BEL and SRIN. Also, the difference between its score and that of the winner was tiny for other test functions. Hence, the JG obtained good diversity performance based on these results.

5.7.2.3 Diversity Evaluation Using Extreme Nondominated Solution Generation

Tables 5.9 and 5.10 list the total number of extreme nondominated solutions found by various MOEAs for different unconstrained and constrained test functions, respectively. For each test function, 50 nondominated solution sets were obtained after 50 simulation runs for each MOEA.

Furthermore, an initial value of a scaling factor for each test function was arbitrarily chosen as $\mu = 0.01$ according to Equations (5.11) and (5.12). If no extreme nondominated solution was counted in all the MOEAs, comparison could not be made. In this case, the value of the scaling factor was repetitively increased by a step (0.01) until at least one nonzero value was found in any of the MOEAs.

From Table 5.9, the results indicated that the JG acquired the largest number of extreme nondominated solutions for most unconstrained test functions. Similar outcomes were observed for the constrained test functions, as shown in Table 5.10.

5.7.2.4 Statistical Test Using Binary ε-Indicator

To make a direct comparison in the performance of the algorithms, the binary ε-indicator was adopted. Tables 5.11–5.13 show the statistical

TABLE 5.9

Total Numbers of Extreme Nondominated Solutions for Unconstrained Test Functions

Scaling Factor	Test Function	MOGA	NPGA2	NSGA2	SPEA2	PAES	MICROGA	JG
0.01	SCH	15	43	60	62	**264**	0	72
0.01	FON	16	22	22	17	0	6	**26**
0.01	POL	32	46	82	45	**106**	34	93
0.01	ZIT1 (binary)	9	17	30	20	4	0	**32**
0.03	ZIT2 (binary)	0	0	0	0	5	0	**12**
0.01	ZIT3 (binary)	1	**12**	10	0	10	0	5
0.7	ZIT4 (binary)	59	0	0	26	0	0	**81**
0.01	ZIT6 (binary)	0	114	57	87	**562**	0	145
0.01	ZIT1 (real)	0	0	0	0	0	0	**13**
0.03	ZIT2 (real)	0	0	0	0	0	0	**3**
0.01	ZIT3 (real)	0	0	0	0	3	0	**10**
0.01	ZIT4 (real)	0	0	0	0	0	0	**54**
0.01	ZIT6 (real)	0	0	0	0	0	0	**9**

Note: The best result for each test function is marked in bold.

TABLE 5.10

Total Number of Extreme Nondominated Solutions for Constrained Test Functions

Scaling Factor	Test Function	MOGA	NPGA2	NSGA2	SPEA2	PAES	MICROGA	JG
0.01	DEB	63	56	64	59	71	14	**73**
0.01	BEL	46	53	86	65	13	26	**89**
0.01	SRIN	6	6	29	18	0	16	**38**
0.01	TAN	18	6	21	11	0	6	**26**
0.01	BINH	2	2	**16**	11	0	0	10

Note: The best result for each test function is marked in bold.

TABLE 5.11

Statistical Results of Binary ε-Indicator in Terms of the Number of Occurrences in Three Different Cases for Unconstrained Test Functions with Binary Coding

Test Function	Case	MOGA	NPGA2	NSGA2	SPEA2	PAES	MICROGA
SCH	Case I	**1,541**	738	742	644	513	**2,329**
	Case II	27	89	274	144	378	0
	Case III	932	**1,673**	1,484	1,712	1,609	171
FON	Case I	290	331	190	152	**1,787**	934
	Case II	13	11	116	96	0	0
	Case III	2,197	2,158	2,194	**2,252**	713	1,566
POL	Case I	1,567	1,568	913	1,207	**2,366**	2,064
	Case II	0	0	636	79	0	0
	Case III	933	932	**951**	1,214	134	436
ZIT1 (binary)	Case I	1,948	1,435	1,028	1,592	0	**2,497**
	Case II	108	453	973	432	0	0
	Case III	444	612	499	476	**2,500**	3
ZIT2 (binary)	Case I	2,337	1,307	1,156	1,473	127	**2,500**
	Case II	68	852	1071	713	0	0
	Case III	95	341	273	314	**2,373**	0
ZIT3 (binary)	Case I	2,055	1,006	987	1,664	796	**2,499**
	Case II	149	915	801	394	0	0
	Case III	296	579	712	442	**1,704**	1
ZIT4 (binary)	Case I	1,514	1,608	1,477	1,509	2,372	**2,500**
	Case II	366	167	892	833	8	0
	Case III	620	725	131	158	120	0
ZIT6 (binary)	Case I	2,481	1,489	1,434	1,858	856	**2,500**
	Case II	10	616	731	441	103	0
	Case III	9	395	335	201	**1,541**	0

Source: Data from Chan, T. M., Man, K. F., Kwong, S., Tang, K. S., A jumping gene paradigm for evolutionary multiobjective optimization, *IEEE Transactions on Evolutionary Computation*, 12(2), 143–159, 2008.

Note: The best result for each case is marked in bold.

TABLE 5.12

Statistical Results of Binary ε-Indicator in Terms of the Number of Occurrences in Three Different Cases for Unconstrained Test Functions with Real Number Coding

Test Function	Case	MOGA	NPGA2	NSGA2	SPEA2	PAES	MICROGA
ZIT1 (real)	Case I	**1,669**	**1,411**	**1,543**	**1,314**	291	**1,397**
	Case II	0	0	0	0	21	0
	Case III	831	1,089	957	1,186	**2,188**	1,103
ZIT2 (real)	Case I	**1,634**	**1,674**	**1,707**	**1,951**	1,110	**1,438**
	Case II	0	0	0	0	49	0
	Case III	866	826	793	549	**1,341**	1,062
ZIT3 (real)	Case I	1,245	**1,340**	1,237	**1,273**	416	**1,400**
	Case II	0	2	0	0	14	0
	Case III	**1,255**	1,158	**1,263**	1,227	**2,070**	1,100
ZIT4 (real)	Case I	**1,670**	**2,078**	**1,308**	**2,294**	**1,422**	**2,500**
	Case II	0	0	0	0	0	0
	Case III	830	422	1,192	206	1,078	0
ZIT6 (real)	Case I	**2,496**	**2,492**	**2,487**	**2,491**	**2,440**	**2,500**
	Case II	0	2	2	4	11	0
	Case III	4	6	11	5	49	0

Source: Data from Chan, T. M., Man, K. F., Kwong, S., Tang, K. S., A jumping gene paradigm for evolutionary multiobjective optimization, *IEEE Transactions on Evolutionary Computation*, 12(2), 143–159, 2008.

Note: The best result for each case is marked in bold.

results of the binary ε-indicator in terms of the number of occurrences in three different cases for unconstrained test functions with binary coding, with real number coding, and constrained test functions, respectively. The three cases were specified as follows:

Case I: JG is better than the compared algorithm.

Case II: JG is worse than the compared algorithm.

Case III: They are incomparable.

The effectiveness of the JG against other MOEAs is summarized in Tables 5.14 and 5.15. The JG scored 58 favorable marks (X) and 20 inconclusive marks (blanks) for the unconstrained test functions. For the constrained test functions, the JG scored 17 favorable marks (X) and 13 inconclusive marks (blanks).

Moreover, the JG had better search performance in the ZIT series for unconstrained test functions. This may indicate that JG is more suitable for solving problems with longer chromosome lengths and real-number coding. Particularly, the JG had outstanding performance on ZIT4, which has many different local Pareto-optimal fronts in the search space. JG was able to

TABLE 5.13

Statistical Results of Binary ε-Indicator in Terms of the Number of Occurrences in Three Different Cases for Constrained Test Functions

Test Function	Case	MOGA	NPGA2	NSGA2	SPEA2	PAES	MICROGA
DEB	Case I	**1,476**	**1,679**	841	724	**2,470**	**1,984**
	Case II	174	106	501	585	0	7
	Case III	850	715	**1,158**	**1,191**	30	509
BEL	Case I	**2,329**	**2,341**	1,190	1,775	**2,374**	**2,467**
	Case II	1	0	39	0	0	7
	Case III	170	159	**1,271**	725	126	26
SRIN	Case I	**1,583**	**1,783**	987	1,419	**2,299**	**1,441**
	Case II	471	338	703	465	97	552
	Case III	446	379	810	616	104	507
TAN	Case I	180	169	302	156	**2,047**	528
	Case II	319	359	202	426	0	24
	Case III	**2,001**	**1,972**	**1,996**	**1,918**	453	**1,948**
BINH	Case I	331	470	145	127	**1,889**	595
	Case II	0	0	76	15	0	0
	Case III	**2,169**	**2,030**	**2,279**	**2,358**	611	**1,905**

Source: Data from Chan, T. M., Man, K. F., Kwong, S., Tang, K. S., A jumping gene paradigm for evolutionary multiobjective optimization, *IEEE Transactions on Evolutionary Computation*, 12(2), 143–159, 2008.

Note: The best result for each case is marked in bold.

provide sufficient diversity to prevent trapping in some local Pareto-optimal fronts. In the constrained test functions, the JG also outperformed other MOEAs in DEB, BEL, and SRIN.

In summary, it was demonstrated that the JG had a powerful and effective search ability to seek better sets of nondominated solutions with good convergence and diversity performance. For demonstration, typical sample sets of nondominated solutions found by different MOEAs for the unconstrained and constrained test functions are shown in Figures 5.2–5.4.

The nondominated solution sets for the test functions ZIT1, ZIT2, ZIT3, ZIT4, and ZIT6 (both binary and real-number coding) obtained by a MICROGA (microgenetic algorithm) were not included since they were too far away from the true Pareto-optimal front.

5.7.3 An Experimental Test of Theorems of Equilibrium

As indicated by the two theorems of equilibrium given in Section 4.3, the element in a primary schema competition set S_ξ of order $o(\xi) = n$ will globally asymptotically converge to $1/2^n$ under copy-and-paste or cut-and-paste operations, despite the initial proportion of the schemata in the set.

In theory, this effect can be visualized by monitoring the proportions of schemata. However, due to the huge number of schemata in a practical

TABLE 5.14

Summary of Statistical Results Obtained from the Binary ε-Indicator for Unconstrained Test Functions

Test Function	JG $\underset{F}{>}$ MOGA	JG $\underset{F}{>}$ NPGA2	JG $\underset{F}{>}$ NSGA2	JG $\underset{F}{>}$ SPEA2	JG $\underset{F}{>}$ PAES	JG $\underset{F}{>}$ MICROGA
SCH	X					X
FON					X	
POL	X	X			X	X
ZIT1 (binary)	X	X	X	X		X
ZIT2 (binary)	X	X	X	X		X
ZIT3 (binary)	X	X	X	X		X
ZIT4 (binary)	X	X	X	X	X	X
ZIT5 (binary)	X	X	X	X		X
ZIT6 (binary)	X	X	x	x		X
ZIT1 (real)	X	X	X	X		X
ZIT2 (real)	X	X	X	X		X
ZIT3 (real)		X		X		X
ZIT4 (real)	X	X	X	X	X	X
ZIT6 (real)	X	X	X	X	X	X

Source: Data from Chan, T. M., Man, K. F., Kwong, S., Tang, K. S., A jumping gene paradigm for evolutionary multiobjective optimization, *IEEE Transactions on Evolutionary Computation*, 12(2), 143–159, 2008.

Note: $\underset{F}{>}$ means "is more favorable than."

design problem, only low-order schemata can be data logged. Yet, with such limitation, it is still possible to visualize this effect. In the following, this is demonstrated by applying a multiobjective genetic algorithm (MOGA) with or without JG onto a practical design example to demonstrate a full case.

TABLE 5.15

Summary of Statistical Results Obtained from the Binary ε-Indicator for Constrained Test Functions

Test Function	JG $\underset{F}{>}$ MOGA	JG $\underset{F}{>}$ NPGA2	JG $\underset{F}{>}$ NSGA2	JG $\underset{F}{>}$ SPEA2	JG $\underset{F}{>}$ PAES	JG $\underset{F}{>}$ MICROGA
DEB	X	X			X	X
BEL	X	X		X	X	X
SRIN	X	X	X	X	X	X
TAN					X	
BINH					X	

Source: Data from Chan, T. M., Man, K. F., Kwong, S., Tang, K. S., A jumping gene paradigm for evolutionary multiobjective optimization, *IEEE Transactions on Evolutionary Computation*, 12(2), 143–159, 2008.

Note: $\underset{F}{>}$ means "is more favorable than."

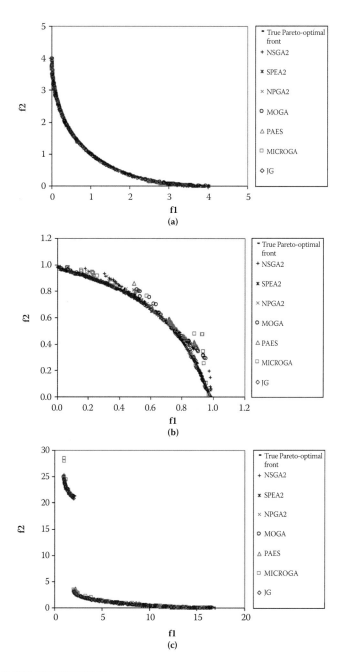

FIGURE 5.2 (SEE COLOR INSERT.)
Sample nondominated solution sets for unconstrained test functions with binary coding. (a) SCH; (b) FON; (c) POL; (d) ZIT1 (binary code); (e) ZIT2 (binary code); (f) ZIT3 (binary code); (g) ZIT4 (binary code); (h) ZIT6 (binary code).

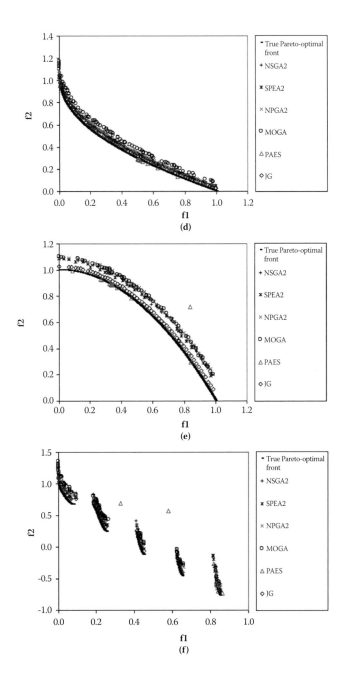

FIGURE 5.2 (CONTINUED)

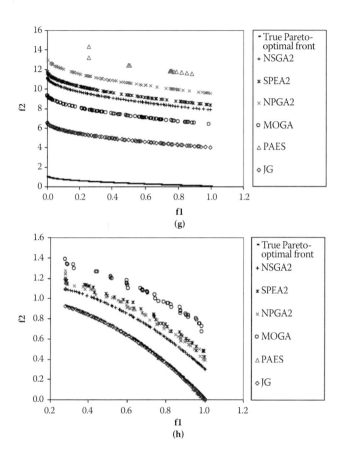

FIGURE 5.2 (CONTINUED)

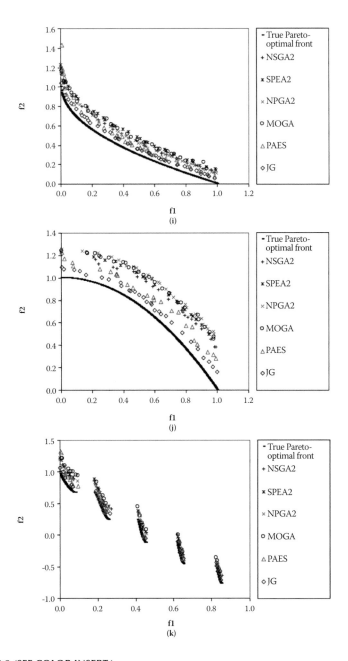

FIGURE 5.3 (SEE COLOR INSERT.)
Sample nondominated solution sets for unconstrained test functions with real-number coding.
(i) ZIT1 (real code); (j) ZIT2 (real code); (k) ZIT3 (real code); (l) ZIT4 (real code); (m) ZIT6 (real code).

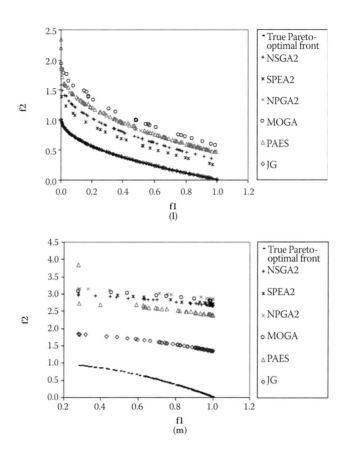

FIGURE 5.3 (CONTINUED)

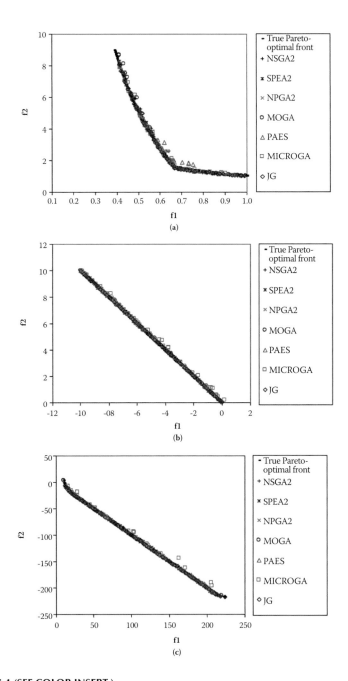

FIGURE 5.4 (SEE COLOR INSERT.)
Sample nondominated solution sets for constrained test functions. (a) DEB; (b) BEL; (c) SRIN; (d) TAN; (e) BINH.

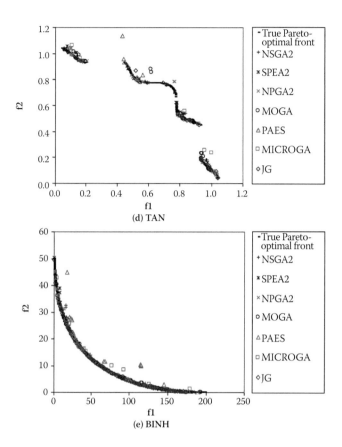

(d) TAN

(e) BINH

FIGURE 5.4 (CONTINUED)

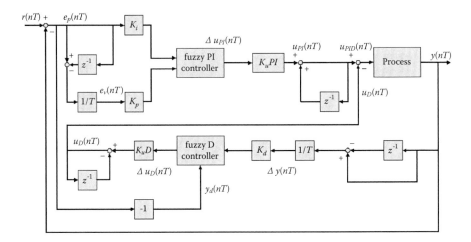

FIGURE 5.5
Fuzzy PID controller system. (From Tang, K. S., Man, K. F., Chen, G., Kwong, S., An optimal fuzzy PID controller, *IEEE Transactions on Industrial Electronics*, 48(4), 757–765, 2001.)

The problem is to design a discrete-time fuzzy proportional-integral-derivative (PID) controller design; its block diagram is depicted in Figure 5.5. Fuzzy PID controllers, evolving from conventional PID controllers, are generally designed for nonlinear systems, higher-order and time-delayed linear systems, and particularly complex and vague systems that cannot be represented by precise mathematical models. The framework of the fuzzy PID control was given in Carvajal, Chen, and Ogmen [3] and Chen [5], and a brief summary is also given in Appendix D. Further details of the design can be found in Misir, Malki, and Chen [12] and Tang, Man, Chen, and Kwong [17].

Although the structure of the fuzzy PID controller given in Figure 5.5 is standard, it remains a difficult task to determine its optimal parameters; this was addressed based on a MOGA [17]. As stated in Fonseca and Fleming [10], instead of having a single solution, a Pareto-optimal solution set can be obtained based on various control performance criteria.

5.7.3.1 *Optimization of Controller Design*

For the design problem, the following time-delayed nonlinear process is assumed:

$$\dot{y}(t) = \sqrt{|y(t)|} + y(t-2) + u_{PID}(t) \tag{5.20}$$

As shown in Figure 5.5 and explained in Appendix D, there are six control parameters to be determined in this design problem. Therefore, the chromosome can be constructed as

$$I = \left\{ K_p, K_i, K_d, R, K_{uD}, K_{uPI} \right\}$$

where each parameter is coded in a 16-bit binary number.

For a general control problem, it is desirable to obtain the optimal results for different system performance. Consider a step input $r(t)$ and the output response $y(t)$; the task is to optimize the following objectives:

1. Minimize the maximum overshoot of the output

$$Obj_1 = \max_t y(t)$$

2. Minimize the settling time of the output

$$Obj_2 = t_s, \text{ such that } 0.98r \leq y(t) \leq 1.02r, \forall t \geq t_s$$

3. Minimize the rising time of the output

$$Obj_3 = t_r = t_2 - t_1, \text{ such that } y(t_1) = 0.1r \text{ and } y(t_2) = 0.9r .$$

5.7.3.2 Results and Comparisons

Referring to the simulation cases given in Section 4.5.4, the primary schemata competition (PSC) sets cannot reach their equilibria in a population with limited size. Instead, the proportions of schemata fluctuate around the equilibrium point, and it is known that the deviation caused by the cut-and-paste operation is smaller than that caused by the copy-and-paste operation.

MOGAs with and without JG were compared; 500 iterations were run for both algorithms, and the population size was fixed at 300. The chromosome length was $L = 96$; the number of order n PSC sets was $\binom{L}{n}$ for $n = 1, 2, \cdots, L$.

Since it is impossible to trace the high-order sets due to their huge size, only order 1 and order 2 PSC sets were analyzed. Figures 5.6 and 5.7 show the mean absolute errors from the equilibria of all order 1 and order 2 PSC sets, respectively. Their means and variances over the last 100 generations were calculated and are shown in Table 5.16. It can be easily noted that the deviation from equilibrium for the cut-and-paste operation was the smallest, which agrees with the results presented in Section 4.5.4; that is, a more even distribution of schemata was promoted using the JG in a GA.

Figure 5.8 shows the obtained set of all rank 1 solutions for the control problem; Figure 5.9 gives the projections on different two-dimensional (2D) planes. Whether the JG is embedded in an MOGA or not, the MOGA can always find some optimal solutions, which demonstrates the great searching ability of the MOGA. However, the effect of using the JG is obvious. Additional different optimal solutions can be located by embedding the JG into the MOGA, which gives a rich set of solutions, thus enhancing the diversity of the solutions.

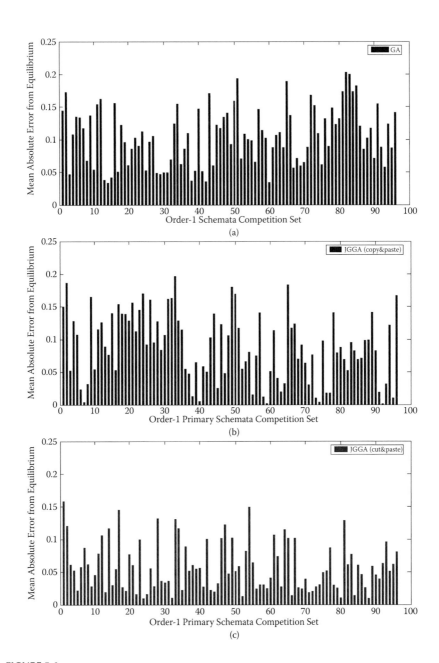

FIGURE 5.6
Mean absolute error from equilibrium of all order-1 primary schemata competition sets in a (a) GA, (b) JG GA using copy-and-paste, and (c) JG GA using cut-and-paste.

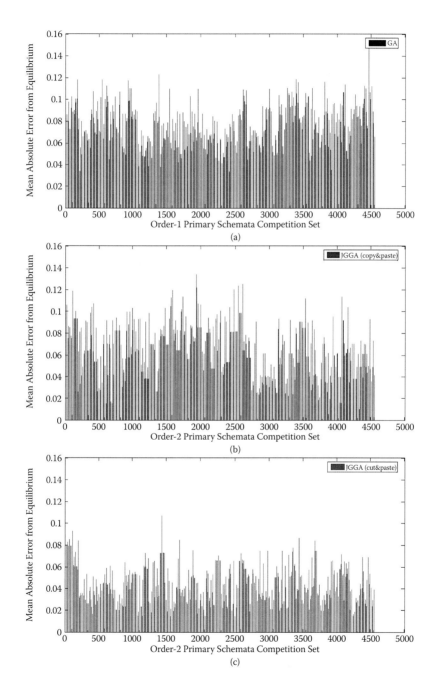

FIGURE 5.7
Mean absolute error from equilibrium of all order-2 primary schemata competition sets in a (a) GA, (b) JG GA using copy-and-paste, and (c) JG GA using cut-and-paste.

TABLE 5.16

Mean Absolute Error from Equilibrium of all Order-1 and Order-2 Primary Schemata Competition Sets

	GA	JGGA (Copy-and-Paste)	JGGA (Cut-and-Paste)
Order-1 PSC set			
Mean	0.0951	0.0888	0.0736
Variance	0.0025	0.0026	0.0024
Order-2 PSC set			
Mean	0.0739	0.0664	0.0560
Variance	0.0006	0.0006	0.0005

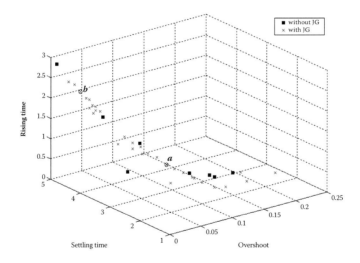

FIGURE 5.8
Obtained Pareto set of solutions. (From Yin, J. J., Tang, K. S., Man K. F., Jumping gene genetic algorithm and its applications in controller design, in *Proceedings of IEEE International Symposium on Industrial Electronics*, Cambridge, UK, June 30–July 2, 2008, 1219–1224.)

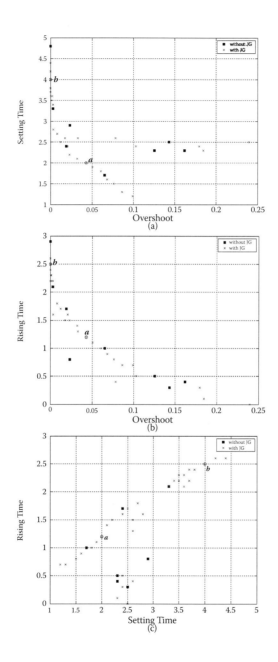

FIGURE 5.9

A 2D projection on (a) overshoot and setting time plane, (b) overshoot and rising time plane, and (c) setting time and rising time plane based on the solutions in Figure 5.6. (From Yin, J. J., Tang, K. S., Man K. F., Jumping gene genetic algorithm and its applications in controller design, in *Proceedings of IEEE International Symposium on Industrial Electronics*, Cambridge, UK, June 30–July 2, 2008, 1219–1224.)

References

1. Belegundu, A. D., Murthy, D. V., Salagame, R. R., Constants, E. W., Multi-objective optimization of laminated ceramic composites using genetic algo-rithms, in *Proceedings of the Fifth AIAA/NASA/USAF/ISSMO Symposium on Multidisciplinary Analysis and Optimization*, Panama City Beach, FL, September 1994, 1015–1022.

2. Binh, T. T., Korn, U., MOBES: A multiobjective evolution strategy for constrained optimization problems, in *Proceedings of Third International Mendel Conference on Genetic Algorithms*, Brno, Czech Republic, June 1997, 176–182.

3. Carvajal, J., Chen, G., Ogmen, H., Fuzzy PID controller: Design, performance evaluation, and stability analysis, *Information Sciences*, 123, 249–270, 2000.

4. Chan, T. M., Man, K. F., Kwong, S., Tang, K. S., A jumping gene paradigm for evolutionary multiobjective optimization, *IEEE Transactions on Evolutionary Computation*, 12(2), 143–159, 2008.

5. Chen, G., Conventional and fuzzy PID controllers: An overview, *International Journal of Intelligent Control Systems*, 1, 235–246, 1996.

6. Coello Coello, C. A., Van Veldhuizen, D. A., Lamont, G. B., *Evolutionary Algorithms for Solving Multi-objective Problems*, New York: Kluwer, 2002.

7. Deb, K., *Multi-objective Optimization Using Evolutionary Algorithms*, Chichester, UK: Wiley, 2001.

8. Deb, K., Jain, S., *Running Performance Metrics for Evolutionary Multi-objective Optimization*, Technical Report (KanGAL Report No. 2002004), Kanpur, India: Kanpur Genetic Algorithms Laboratory, Indian Institute of Technology, 2002.

9. Deb, K., Pratap, A., Agrawal, S., Meyarivan, T., A fast and elitist multiobjective genetic algorithm: NSGA-II, *IEEE Transactions on Evolutionary Computation*, 6(2), 182–197, 2002.

10. Fonseca, C. M., Fleming, P. J., Multiobjective optimization and multiple con-straint handling with evolutionary algorithms—Part II: Application example, *IEEE Transactions on System, Man and Cybernetic Part A: Systems and Humans*, 28(1), 38–47, 1998.

11. Miller, W. J., McDonald, J. F., Pinsker, W. Molecular domestication of mobile ele-ments, *Genetica*, 100(1–3), 261–270, 1997.

12. Misir, D., Malki, H. A., Chen, G., Design and analysis of a fuzzy proportional-integral-derivative controller, *International Journal of Fuzzy Sets and Systems*, 79, 297–314, 1996.

13. Poloni, C., Giurgevich, A., Onesti, L., Pediroda, V., Hybridization of a multi-objective genetic algorithm, a neural network and a classical optimizer for a complex design problem in fluid dynamics, *Computer Methods in Applied Mechanics and Engineering*, 186(2–4), 403–420, 2000.

14. Schaffer, J. D., Multiple objective optimization with vector evaluated genetic algorithms, in *Proceedings of the First International Conference on Genetic Algorithms and Their Applications*, Pittsburgh, July 1985, 93–100.

15. Srinivas, N., Deb, K., Multiobjective function optimization using nondominated sorting genetic algorithms, *Evolutionary Computation*, 2(3), 221–248, 1994.

16. Tanaka, M., Watanabe, H., Furukawa, Y., Tanino, T., GA-based decision support system for multicriteria optimization, in *Proceedings of IEEE International Conference on Systems, Man and Cybernetics*, Vancouver, Canada, October 1995, 2:1556–1561.

17. Tang, K. S., Man, K. F., Chen, G., Kwong, S., An optimal fuzzy PID controller, *IEEE Transactions on Industrial Electronics*, 48(4), 757–765, 2001.

18. Veldhuizen, D. A. V., Multiobjective evolutionary algorithms: classifications, analyses, and new innovations, Ph.D. dissertation, Department of Electrical and Computer Engineering, Graduate School of Engineering, Air Force Institute of Technology, Wright Patterson AFB, OH, May 1999.

19. Yin, J. J., Tang, K. S., Man K. F., Jumping gene genetic algorithm and its applications in controller design, in *Proceedings of IEEE International Symposium on Industrial Electronics*, Cambridge, UK, June 30–July 2, 2008, 1219–1224.

20. Zitzler, E., Evolutionary algorithms for multiobjective optimization: Methods and applications, Ph.D. dissertation, Swiss Federal Institute of Technology (ETH), Zurich, Switzerland, November 1999.

21. Zitzler, E., Deb, K., Thiele, L., Comparison of multiobjective evolutionary algorithms: Empirical results, *Evolutionary Computation*, 8(2), 173–195, 2000.

22. Zitzler, E., Laumanns, M., Thiele, L., Fonseca, C. M., Grunert da Fonseca, V., Why quality assessment of multiobjective optimizers is difficult, in *Proceedings of the Genetic and Evolutionary Computation Conference*, New York, July 2002, 666–674.

23. Zitzler, E., Thiele, L., Laumanns, M., Fonseca, C. M., Grunert da Fonseca, V., Performance assessment of multiobjective optimizers: An analysis and review, *IEEE Transactions on Evolutionary Computation*, 7(2), 117–132, 2003.

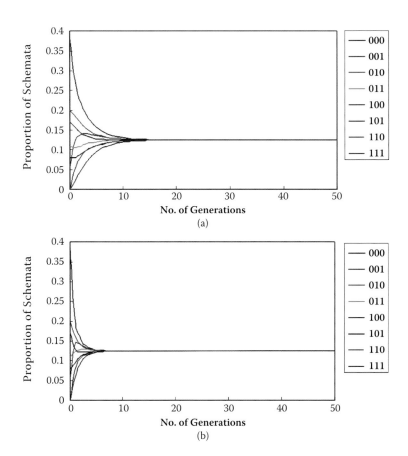

COLOR FIGURE 4.4
The proportion of schemata against generations using (a) copy-and-paste and (b) cut-and-paste operations, where 000 stands for ***0*0***0****** and so on. (From Yin, J. J., Yeung, S. H., Tang, W. K. S., Man, K. S., Kwong, S., Enhancement of multiobjective search: A Jumping-genes approach, in *Proceedings of IEEE International Symposium on Industrial Electronics*, Vigo, Spain, June 2007, 1855–1858.)

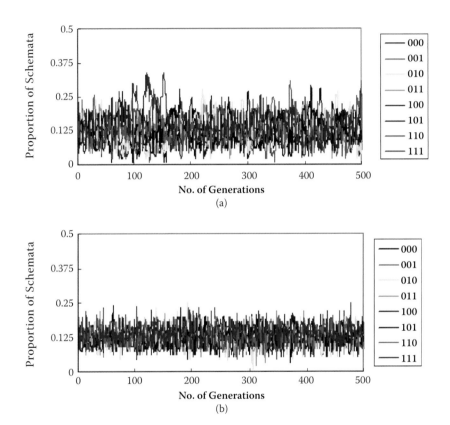

COLOR FIGURE 4.6
The proportion of schemata against generations in a finite population using (a) copy-and-paste and (b) cut-and-paste operations. (From Tang, K. S., Yin, R. J., Kwong, S., Ng, K. T., Man, K. F., A theoretical development and analysis of jumping gene genetic algorithm, *IEEE Transactions on Industrial Informatics*, 7(3), 2011, 408–418.)

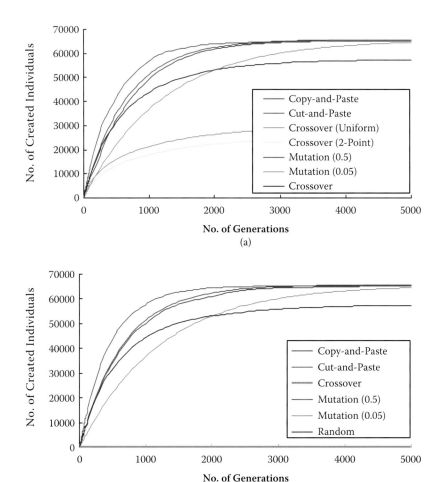

COLOR FIGURE 4.7
The searching abilities of different operations in a finite population with initial population (a) randomly generated and (b) of identical chromosomes. (From Tang, K. S., Yin, R. J., Kwong, S., Ng, K. T., Man, K. F., A theoretical development and analysis of jumping gene genetic algorithm, *IEEE Transactions on Industrial Informatics*, 7(3), 2011, 408–418.)

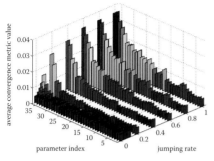

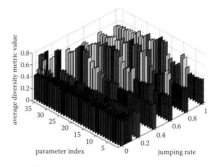

Performance on convergence   Performance on diversity

Unconstrained test function: FON

(a)

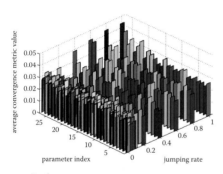

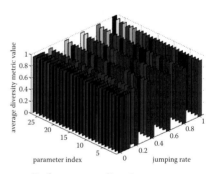

Performance on convergence   Performance on diversity

Unconstrained test function: POL

(b)

COLOR FIGURE 5.1

The 3D plots for average values of convergence and diversity metrics against different parameter indexes and jumping rates for different test functions. (a) Unconstrained test function: FON; (b) Unconstrained test function: POL; (c) Constrained test function: SRIN; (d) constrained test function: TAN. (From Chan, T. M., Man, K. F., Kwong, S., Tang, K. S., A jumping gene paradigm for evolutionary multiobjective optimization, *IEEE Transactions on Evolutionary Computation*, 12(2), 143–159, 2008.)

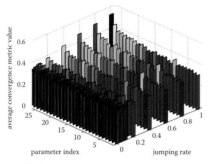

Performance on convergence

Performance on diversity

Constrained test function: SRIN

(c)

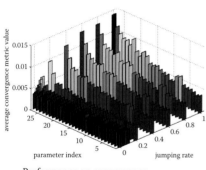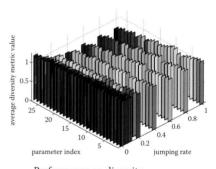

Performance on convergence

Performance on diversity

Constrained test function: TAN

(d)

COLOR FIGURE 5.1 (CONTINUED)

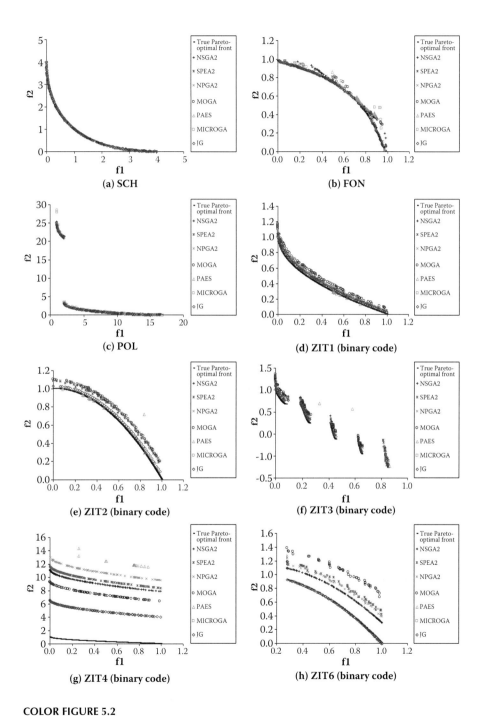

COLOR FIGURE 5.2
Sample nondominated solution sets for unconstrained test functions with binary coding. (a) SCH; (b) FON; (c) POL; (d) ZIT1 (binary code); (e) ZIT2 (binary code); (f) ZIT3 (binary code); (g) ZIT4 (binary code); (h) ZIT6 (binary code).

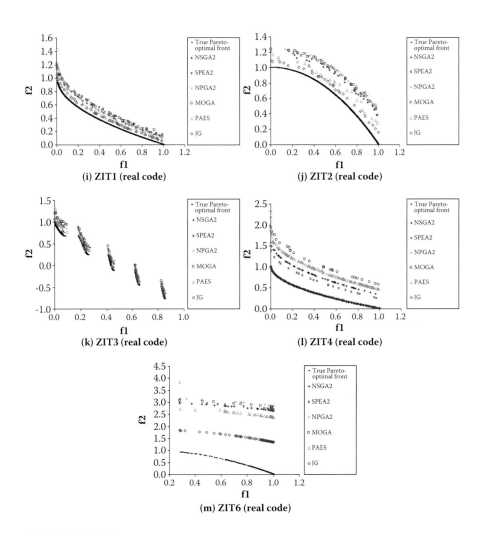

COLOR FIGURE 5.3
Sample nondominated solution sets for unconstrained test functions with real-number coding.
(i) ZIT1 (real code); (j) ZIT2 (real code); (k) ZIT3 (real code); (l) ZIT4 (real code); (m) ZIT6 (real code).

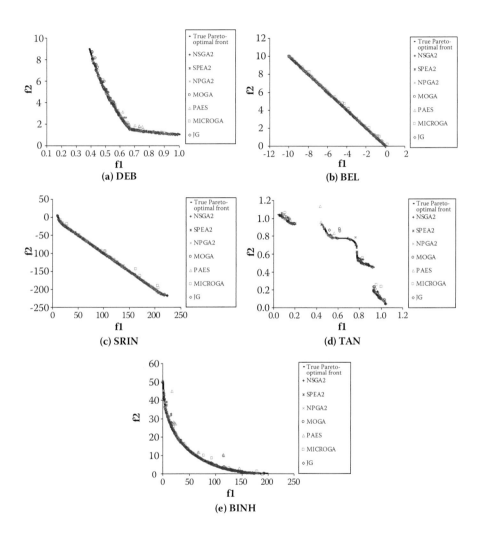

COLOR FIGURE 5.4
Sample nondominated solution sets for constrained test functions. (a) DEB; (b) BEL; (c) SRIN; (d) TAN; (e) BINH.

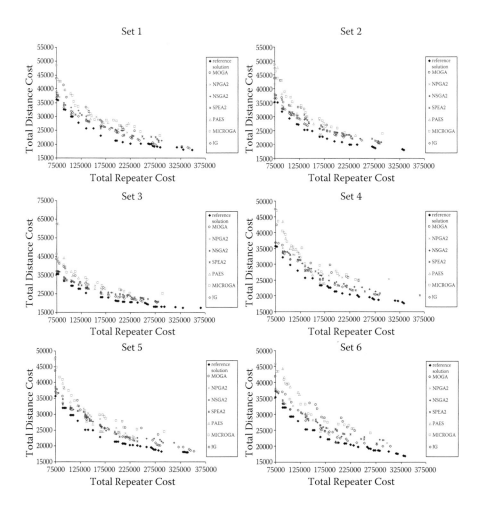

COLOR FIGURE 6.6
Sample nondominated solution sets obtained from each MOEA for 10 sets of testing scenarios.

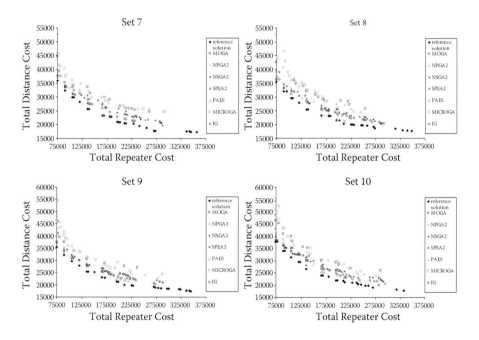

COLOR FIGURE 6.6 (CONTINUED)

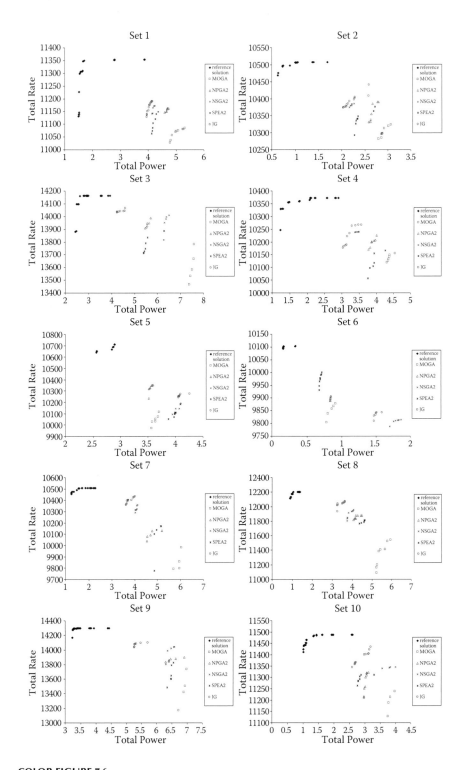

COLOR FIGURE 7.6
Sampled nondominated solution sets obtained by different MOGAs for scenario a: 25 users.

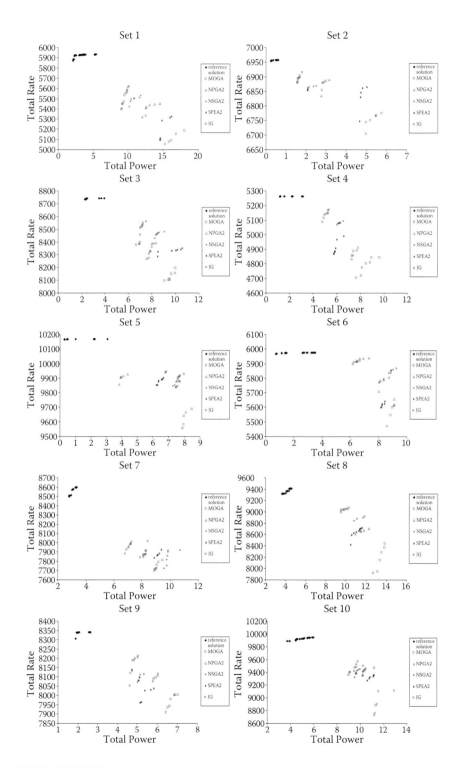

COLOR FIGURE 7.7
Sampled nondominated solution sets obtained by different MOGAs for scenario b: 50 users.

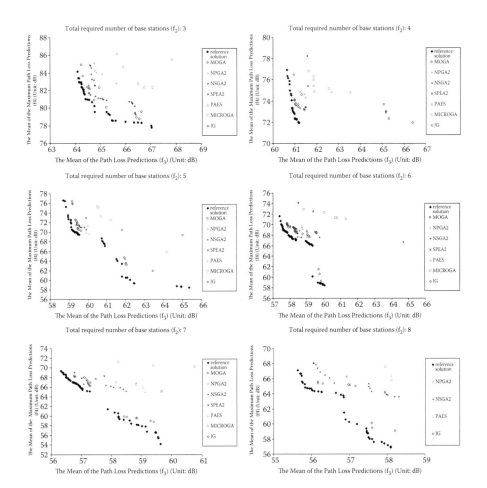

COLOR FIGURE 8.6
Distribution of reference solutions and nondominated solutions obtained by MOGAs with different numbers of base stations for scenario a: 90 dB.

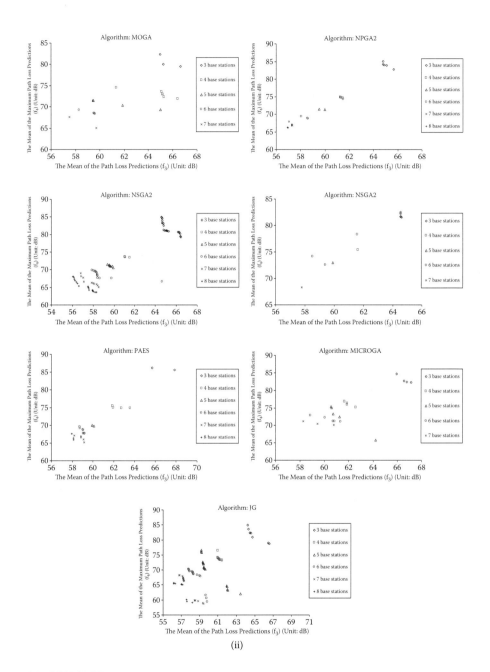

COLOR FIGURE 8.7
Sample nondominated solution sets obtained by different MOGAs for scenario a: 90 dB.

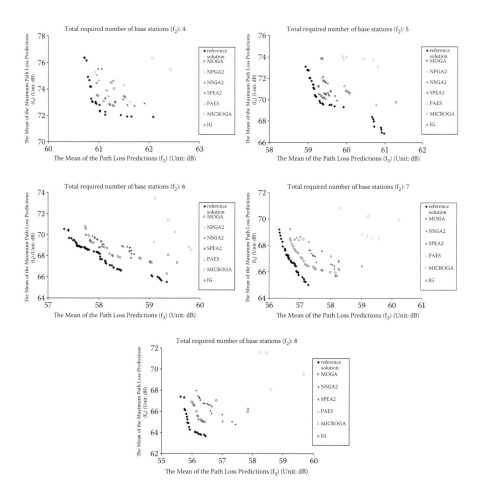

COLOR FIGURE 8.8
Distribution of reference solutions and nondominated solutions obtained by MOGAs with different numbers of base stations for scenario a: 80 dB. (From Chan, T. M., Man, K. F., Tang, K. S., Kwong, S., A jumping-genes paradigm for optimizing factory WLAN network, *IEEE Transactions on Industrial Informatics*, 3(1), 33–43, 2007.)

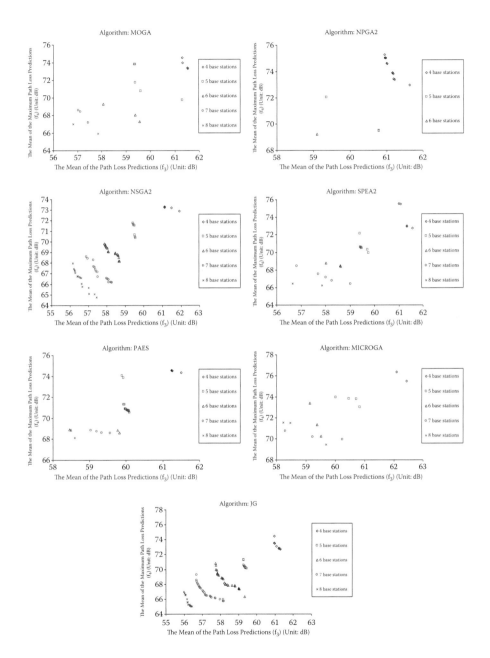

COLOR FIGURE 8.9

Sample nondominated solution sets obtained by different MOGAs for scenario a: 80 dB. (From Chan, T. M., Man, K. F., Tang, K. S., Kwong, S., A jumping-genes paradigm for optimizing factory WLAN network, *IEEE Transactions on Industrial Informatics*, 3(1), 33–43, 2007.)

6

Radio-to-Fiber Repeater Placement in Wireless Local-Loop Systems

6.1 Introduction

Fixed-wire systems (e.g., telephone twisted pairs and coaxial cables) are no longer the only option for providing telephone services. With the emergence of a new technology, the wireless local-loop (WLL) system is now possible for telephone networks.

The WLL system employs wireless links to connect subscribers (equipment such as telephones, fax machines, etc.) to the local exchanges in the central office [5]. It offers services that include voice, fax, data, and Integrated Services Digital Network (ISDN) services, and so on, with the same quality of service in bit error ratio, delay time, and blocking probability as those provided in fixed-wire systems [7,18,19].

The WLL system gives a telecommunication infrastructure that not only better serves the demands of low population densities such as in urban/suburban or rural areas [10, 22] but also is a good solution to alleviate the difficulties of providing telephone services in isolated areas and regions subject to natural disasters such as earthquakes, floods, and hurricanes [7].

There are two types of WLL systems: narrowband WLL systems operating under 10 GHz and broadband WLL systems operating above 10 GHz [6]. In this study, only the narrowband is considered. Different technologies, such as cordless systems, second-generation cellular systems, satellite systems, fixed wireless access systems, and so on, can be utilized in WLL systems [2,11,12,15]. However, most of the narrowband WLL systems are developed using cordless and second-generation cellular systems [18].

WLL systems have a number of advantages over fixed-wire systems from the perspective of service providers and subscribers [3,7,8,16,18,20,21], including

1. faster deployment;
2. low installation cost;

3. low maintenance cost;

4. low network extension cost; and

5. high system capacity.

Moreover, due to the reduction of electronic costs, WLL systems are becoming increasingly competitive. This approach arouses much interest in developing countries, such as those in Latin America, Africa, and the Middle East, as well as in China, where the infrastructures are still inadequate for basic telephone services or the high cost of fixed-wire systems is not preferable [9,21], and the market is huge. Indeed, due to the merits of WLL systems, they have already been installed in some countries, for example, in Japan, Egypt, Turkey, Italy, Spain, and the United Kingdom [3,6,14,20,21].

Despite the obvious advantages, there exists a severe problem in WLL systems: the phenomenon of weak diffraction revealed by field measurement at the 2.3-GHz frequency band [23]. This introduces many noncoverage holes within the cell. An average diffraction loss related to various types of buildings can be up to 21 dB [1]. To resolve this problem, radio-to-fiber repeaters are commonly adopted [17].

The radio-to-fiber repeater uses two types of transmission media. The signals between a base station and a repeater are transmitted through the wireless channel, while those between a repeater and a terminal are transmitted through optical fiber. The main function of the radio-to-fiber repeater is to transmit/receive radio signals to/from base stations and the signals to/from terminals through the optical fiber. Unlike conventional WLL systems, the total system cost for this option depends on the number of additional repeaters installed and the optical fibers (links).

Figure 6.1 depicts the WLL cell with repeaters and links. The total cost consists of the connection cost (between WLL terminals and repeaters) and the repeater cost, which is closely related to repeater antenna gain. The longer the distances between WLL terminals and repeaters are, the higher the total connection cost is. Also, the larger the gain of the repeater antennas used, the higher the total repeater cost. Thus, the locations of radio repeaters should be correctly determined to minimize the total repeater cost and total connection cost simultaneously.

This causes an optimization problem. Consider a schematic diagram for a WLL cell with 10,000 grids, as illustrated in Figure 6.2, where each grid refers to an area 10×10 m^2.

This optimization problem was considered a single-objective optimization problem [17] in which the objective is to minimize the sum of the total repeater cost and total link cost. However, since these two costs conflict, it is more appropriate to handle it in a multiobjective optimization manner [4].

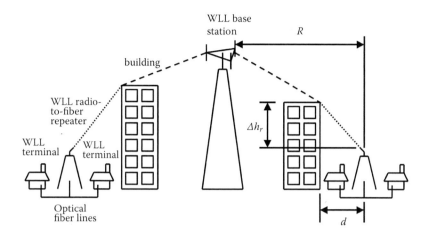

FIGURE 6.1
A WLL cell containing repeaters and links. (From Chan, T. M., Man, K. F., Tang, K. S., Kwong, S., Multiobjective optimization of radio-to-fiber repeater placement using a jumping gene algorithm, in *Proceedings of IEEE International Conference on Industrial Technology*, Hong Kong, China, December 2005, 291–296.)

1	2	3	· · · · · ·	50	51	· · · · · ·	98	99	100
101	102	103	· · · · · ·	150	151	· · · · · ·	198	199	200
201	202	203	· · · · · ·	250	251	· · · · · ·	298	299	300
⋮	⋮	⋮	⋱	⋮	⋮	⋰	⋮	⋮	⋮
4901	4902	4903	· · · · · ·	4950	4951	· · · · · ·	4998	4999	5000
5001	5002	5003	· · · · · ·	5050	5051	· · · · · ·	5098	5099	5100
⋮	⋮	⋮	⋰	⋮	⋮	⋱	⋮	⋮	⋮
9701	9702	9703	· · · · · ·	9750	9751	· · · · · ·	9798	9799	9800
9801	9802	9803	· · · · · ·	9850	9951	· · · · · ·	9898	9899	9900
9901	9902	9903	· · · · · ·	9950	9951	· · · · · ·	9998	9999	10000

Base Station

FIGURE 6.2
A WLL cell comprised of 10,000 grids. (From Chan, T. M., Man, K. F., Tang, K. S., Kwong, S., Multiobjective optimization of radio-to-fiber repeater placement using a jumping gene algorithm, in *Proceedings of IEEE International Conference on Industrial Technology*, Hong Kong, China, December 2005, 291–296.)

6.2 Path Loss Model

To evaluate the total path loss and satisfy the coverage performance, a path loss model, based on the Joint Technical Committee (JTC) model with some modifications, was employed to adapt for the WLL system [23]. In this model, the power loss caused by multiple diffractions from rows of building can be neglected because a directional antenna is used as the WLL repeater antenna. This type of antennae is capable of improving the received signal strength.

Since WLL systems are established for fixed users, there is no mobility effect. Assuming that the WLL base station antenna is located above the rooftop levels of all buildings (see Figure 6.1), the total path loss is expressed as

$$L = L_1 + L_2 \tag{6.1}$$

with the path loss due to free space L_1 and the diffraction loss from rooftop to street L_2 given by

$$L_1 = 20 \log \left(\frac{\lambda}{4 \pi R} \right) \tag{6.2}$$

and

$$L_2 = 10 \log \left[\frac{\lambda}{2 \pi^2 r} \left(\frac{1}{\theta} - \frac{1}{2 \pi + \theta} \right)^2 \right] \tag{6.3}$$

where $\theta = \tan^{-1} \left(\frac{\Delta h_r}{d} \right)$ and $r = \sqrt{\left(\Delta h_r \right)^2 + d^2}$, λ is the wavelength, R is the distance between the WLL base station and the repeater, d is the horizontal distance between the WLL repeater and the diffraction edge, and Δh_r is the height difference between the rooftop level and the WLL repeater antenna.

The paths of radio propagation can be seen in Figure 6.1. To satisfy the coverage performance of each grid,

$$P_0 + G_b + G_r - L \geq P_r \tag{6.4}$$

where P_0 is the WLL base station transmission power (dBm), G_b is the WLL base station antenna gain (dBi), G_r is the WLL repeater antenna gain used on the grid considered (dBi), L is the total path loss (dB), and P_r is the required minimum received signal strength at the WLL repeater (dBm).

TABLE 6.1

List of Notations

Notation	Meaning
N	A set of grids $\{1,2,3,\ldots,|N|\}$
T	A set of terminals $\{1,2,3,\ldots,|T|\}$
d_{ij}	The distance connection cost between WLL terminal i and the repeater installed on grid j $\forall i \in T, \forall j \in N$
C_r	The installation cost of the repeater on each grid
α	The cost coefficient of the repeater antenna with respect to its gain
$(G_r)_j$	The antenna gain of the repeater using on grid j $\forall j \in N$
C_{max}	The maximum capacity of each repeater
P_0	The transmission power of the WLL base station
G_b	The antenna gain of the WLL base station
L_j	The total path loss on grid j $\forall j \in N$
P_r	The required minimum received signal strength at the WLL repeater
R_{min}	The minimum allowed total number of repeaters to be installed
R_{max}	The maximum allowed total number of repeaters to be installed
x_{ij}	Equal to 1 if WLL terminal i is connected to the repeater installed on grid j and 0 otherwise $\forall i \in T, \forall j \in N$
y_j	Equal to 1 if a repeater is installed on grid j and 0 otherwise $\forall j \in N$
z_j	Equal to 1 if a terminal is located on grid j and 0 otherwise $\forall j \in N$

Source: Data from Chan, T. M., Man, K. F., Tang, K. S., Kwong, S., Multiobjective optimization of radio-to-fiber repeater placement using a jumping gene algorithm, in *Proceedings of IEEE International Conference on Industrial Technology*, Hong Kong, China, December 2005, 291–296.)

6.3 Mathematical Formulation

In this repeater placement problem, the objective was to minimize the total repeater cost and total link cost simultaneously. The mathematical model in Park, Song, and Bae [17] was adopted, and the notations used are given in Table 6.1.

The objective functions were specified by two types of costs: the total repeater cost, including the cost of repeater installation and repeater antennas, and the total link cost between WLL terminals and repeaters, which was calculated as the Euclidean distance between a WLL terminal and a repeater. In mathematics, minimize the following two functions:

$$\text{Total repeater cost: } f_1 = \sum_{j \in N}\left(C_r + \alpha \cdot (G_r)_j\right) \cdot y_j \tag{6.5}$$

$$\text{Total link cost: } f_2 = \sum_{i \in T}\sum_{j \in N} d_{ij} x_{ij} \tag{6.6}$$

subject to the following constraints:

1. Each WLL terminal must be connected to only one repeater:

$$\sum_{j \in N} x_{ij} = 1, \quad \forall i \in T$$

2. The maximum number of WLL terminals c_{max} that can be connected to one repeater:

$$\sum_{i \in T} x_{ij} \leq c_{max} \cdot y_j, \quad \forall j \in N$$

3. The value of the received signal strength at the repeater on grid j must be larger than or equal to P_r to satisfy the coverage performance $\forall j \in N$:

$$(P_0 + G_b + (G_r)_j - L_j) \cdot y_j \geq P_r, \quad \forall j \in N$$

4. The total number of repeaters to be installed must be within the range of $[R_{min}, R_{max}]$:

$$R_{min} \leq \sum_{j \in N} y_j \leq R_{max}$$

5. The total number of terminals $|T|$ supported cannot exceed the sum of maximum capacity of all installed repeaters:

$$C_{max} \cdot \sum_{j \in N} y_j \geq |T|$$

6. A repeater cannot be installed on grid j where a terminal is located $\forall j \in N$:

$$y_j + z_j \leq 1, \quad \forall j \in N$$

7. The integer variables x_{ij}, y_j, and z_j are limited to be 0 or 1 only $\forall i \in T, \forall j \in N$:

$$x_{ij}, y_j, z_j \in \{0,1\}, \quad \forall i \in T, \forall j \in N$$

6.4 Chromosome Representation

To provide further design flexibility, a hierarchical chromosome structure, which was first proposed in Man, Tang, and Kwong [13], was employed. A hierarchical chromosome consists of two types of genes, control genes and parameter genes; each control gene is a bit, while each parameter gene is an integer number representing the selected grid number on which a repeater will be installed (as shown in Figure 6.2). The bit of a control gene determines whether the corresponding repeater is installed. Duplicate grid numbers are not allowed in the same chromosome because only one repeater can be installed on each grid.

This encoding scheme allows searching for a different number of repeaters. Assuming that the total number of repeaters to be installed can vary from 5 to 25, an example of the corresponding chromosome is illustrated in Figure 6.3. The hierarchical chromosome length should be 50 (25 control genes plus 25 parameter genes). If the ith control gene's value is 1, the corresponding grid number in the $(i + 25)$th gene (i.e., the ith parameter gene) is chosen for repeater installation and is 0 otherwise, where $i = 1, 2, \cdots, 25$.

Control gene no.	Control gene	Parameter gene	Description
1	1	5235	Grid number, 5235, is selected for installing a repeater
2	1	1315	Grid number, 1315, is selected for installing a repeater
3	0	64	Grid number, 64, is **not** selected
⋮	⋮	⋮	⋮
25	1	6154	Grid number, 6154, is selected for installing a repeater

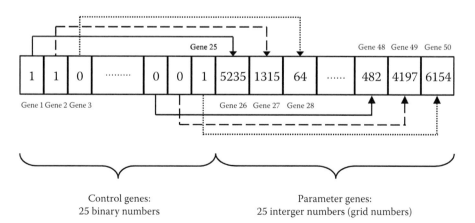

FIGURE 6.3
Encoding method for a hierarchical chromosome. (From Chan, T. M., Man, K. F., Tang, K. S., Kwong, S., Multiobjective optimization of radio-to-fiber repeater placement using a jumping gene algorithm, in *Proceedings of IEEE International Conference on Industrial Technology*, Hong Kong, China, December 2005, 291–296.)

Referring to the example given in Figure 6.3, the grid numbers 5,235, 1,315, and 6,154 were selected for repeater installation since the corresponding control genes were 1. On the other hand, the grid numbers 64, 482, and 4,197 were not chosen because the corresponding control genes were 0.

6.5 Jumping Gene Transposition

As there are two types of genes in our chromosome-encoding method, two corresponding types of transposons can exist in each chromosome. For ease in referencing, they are called control transposons and parameter transposons. Control transposons only present and jump in the range from the 1st gene to the 25th gene, as depicted in Figure 6.3, while parameter transposons are between the 26th and 50th genes. A chromosome will be infeasible if it jumps to any position outside its own specified ranges.

As mentioned in Section 6.4, duplicate grid numbers were not permitted in a chromosome since only one repeater could be installed on each grid. As a result, the copy-and-paste transposition was only performed in two different chromosomes.

6.6 Chromosome Repairing

After performing the genetic operations (i.e., crossover and mutation), duplicated integers may emerge in a chromosome; hence an invalid chromosome results. Therefore, a repairing mechanism is implemented so that these invalid chromosomes are repaired. The repairing mechanism is depicted in Figure 6.4. When one or more pairs of repeating integers exist in a chromosome, all

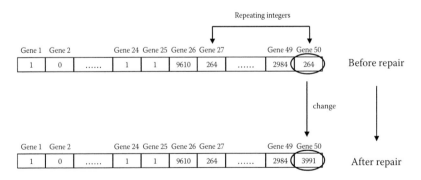

FIGURE 6.4
Repairing a chromosome.

except one of them are to be replaced by a randomly generated integer to resolve the conflict. Referring to the example in Figure 6.4, the chromosome has a pair of repeating values, 264, in genes 27 and 50 originally. To turn it into a valid one, gene 50 has been randomly selected and is replaced by a newly generated integer value, say 3,991.

6.7 Results and Discussion

The parameters of WLL systems and multiobjective evolutionary algorithms (MOEAs) (multiobjective genetic algorithm [MOGA], niched Pareto genetic algorithm 2 [NPGA2], nondominated sorting genetic algorithm 2 [NSGA2], strength Pareto evolutionary algorithm 2 [SPEA2], Pareto archived evolution strategy [PAES], microgenetic algorithm [MICROGA], and jumping gene [JG]) used in the simulations are given in Tables 6.2 and 6.3, respectively.

TABLE 6.2

Parameters of WLL Systems

Parameter	Value
Total number of grids in the service area $|N|$	10,000
Grid size	$10 \times 10 \text{ m}^2$
Total number of terminals $|T|$	200 with uniform distribution on the total number of grids
Range of total allowed number of repeaters R to be installed	5–25
Height difference between the average rooftop level and the WLL repeater antenna Δh_r	Random selected from the interval [1,200] (unit: meter)
Horizontal distance between the WLL repeater and the diffracting edge d	Random selected from the interval [10,200] (unit: meter)
Wavelength λ (frequency)	$3 \times 10^8 \text{ ms}^{-1} (2.35 \times 10^9 \text{ Hz})$
WLL base station transmission power P_0	40 dBm
Received signal strength at WLL repeater P_r	−80 dBm
WLL base station antenna gain G_b	6 dBi
WLL repeater antenna gain $(G_r)_j$	10 dBi for type I: Directional antenna (120°)
	13 dBi for type II: Directional antenna (60°)
	16 dBi for type III: Directional antenna (30°)
Maximum capacity of a repeater C_{max}	8–40
Installation cost of a repeater for any grid C_r	5,000
Cost coefficient of repeater antenna with respect to its gain α	1,000

Source: Data from Chan, T. M., Man, K. F., Tang, K. S., Kwong, S., Multiobjective optimization of radio-to-fiber repeater placement using a jumping gene algorithm, in *Proceedings of IEEE International Conference on Industrial Technology*, Hong Kong, China, December 2005, 291–296.)

TABLE 6.3

Parameters of MOEAs

Parameter	Value/Type
Population size	
MOGA, NPGA2, NSGA2, SPEA2, JG	100
PAES	1
MICROGA	4
Maximum generations or iterations	
MOGA, NPGA2, NSGA2, SPEA2, JG	1,500
PAES	150,000
MICROGA	37,500
Crossover type	Uniform crossover
Crossover rate	0.8
Mutation rate	0.1
Other settings for JG	
Jumping rate	0.01
Number of transposons	2
Length of transposons	5
Other settings for SPEA2 or PAES	
Archive size (SPEA2, PAES)	100
Depth (PAES)	4
Other settings for MICROGA	
Size of external memory	100
Size of population memory	80
Percentage of nonreplaceable memory	0.25
Replacement cycle	Every 25 iterations
Number of subdivisions of the adaptive grid	25
Number of iterations to achieve nominal convergence	4

Source: Data from Chan, T. M., Man, K. F., Tang, K. S., Kwong, S., Multiobjective optimization of radio-to-fiber repeater placement using a jumping gene algorithm, in *Proceedings of IEEE International Conference on Industrial Technology,* Hong Kong, China, December 2005, 291–296.)

In the simulations, 10 sets of different testing scenarios were considered; the locations of the terminals, the height differences between the average rooftop level and the WLL repeater antenna Δh_r, and the horizontal distances between the WLL repeater and the diffracting edge d were randomly generated in each testing set.

To evaluate the quality of the nondominated solution sets found by each MOEA, two performance metrics, the Deb and Jain convergence metric and spread (mentioned in Sections 5.2 and 5.3 of Chapter 5, respectively) were adopted. The means and standard deviations of these two performance metrics were obtained based on 40 simulation runs in every set of testing scenarios.

Since the true Pareto-optimal solutions set is unknown, a set of reference solutions was found to approximate the true Pareto-optimal set for the calculations of the performance metrics. This reference set was obtained by incorporating all the MOEAs together with some arbitrarily but sufficiently large generations or iterations (3,000 generations for MOGA, NPGA2, NSGA2, SPEA2, and JG; 300,000 iterations for PAES; and 75,000 iterations for MICROGA).

6.7.1 Mean and Standard Deviation of Deb and Jain Convergence Metric for Evaluating Convergence

The means and standard deviations of the Deb and Jain convergence metric are given in Table 6.4. Note that smaller means (i.e., better convergence) were obtained by the JG in five testing sets (sets 2, 3, 4, 6, and 10). In contrast, the NSGA2 and SPEA2 had better convergence in four testing sets (sets 1 and 7–9) and one testing set (set 5), respectively.

6.7.2 Mean and Standard Deviation of Spread for Evaluating Diversity

The means and standard deviations of the diversity metric spread are given in Table 6.5. The JG and NSGA2 acquired smaller means (i.e., better diversity) in two testing sets (sets 5 and 7) and one testing set (set 6), respectively. However, the SPEA2 obtained better diversity in seven testing sets (sets 1, 2, 3, 4, and 8–10).

6.7.3 Diversity Evaluation Using Extreme Nondominated Solution Generation

Figure 6.5 illustrates the total number of extreme nondominated solutions found by various MOEAs for all 10 sets of different testing scenarios. The scaling factor, as seen in Equations (5.11) and (5.12), was set as 0.01. In each testing case, 40 simulation runs were carried out for each MOEA. From the figure, the results indicated that the total number of extreme nondominated solutions found by the JG was the largest.

6.7.4 Statistical Test Using Binary ε-Indicator

The statistical results of the binary ε-indicator in terms of the number of occurrences of three comparison cases for the 10 sets of different testing scenarios are presented in Table 6.6; the three comparison cases were specified as follows:

Case I: JG is better than the compared algorithm.
Case II: JG is worse than the compared algorithm.
Case III: They are incomparable.

TABLE 6.4

Means and Standard Deviations of the Deb and Jain Convergence Metric for 10 Sets of Testing Scenarios Using Different Algorithms

Set No.	MOGA	NPGA2	NSGA2	SPEA2	PAES	MICROGA	JG
1	0.18523	0.10565	**0.09517**	0.09543	0.13061	0.22392	0.09572
	(0.02625)	(0.02558)	(0.01077)	**(0.00956)**	(0.04365)	(0.02260)	(0.00978)
2	0.20975	0.11147	0.10622	0.10764	0.14184	0.25765	**0.10448**
	(0.03841)	(0.02740)	(0.01076)	(0.01067)	(0.04286)	(0.02552)	**(0.01059)**
3	0.20242	0.10224	0.09488	0.09666	0.12858	0.21525	**0.09473**
	(0.05163)	(0.01753)	**(0.00973)**	(0.01098)	(0.04167)	(0.02190)	(0.01359)
4	0.19861	0.10580	0.09885	0.09636	0.11902	0.22935	**0.09310**
	(0.03369)	(0.02133)	(0.00921)	**(0.00846)**	(0.03693)	(0.02593)	(0.01176)
5	0.18719	0.10503	0.09922	**0.09536**	0.12918	0.22970	0.09625
	(0.02767)	(0.01591)	(0.01176)	**(0.00847)**	(0.03602)	(0.02161)	(0.00996)
6	0.19051	0.09751	0.09761	0.09702	0.13096	0.22927	**0.09235**
	(0.02317)	(0.01811)	**(0.00752)**	(0.01261)	(0.04187)	(0.01625)	(0.00793)
7	0.19037	0.10617	**0.10091**	0.10336	0.13138	0.22712	0.10169
	(0.02533)	(0.02357)	(0.01148)	**(0.01077)**	(0.04013)	(0.02298)	(0.01129)
8	0.19563	0.10689	**0.09695**	0.09798	0.11998	0.23053	0.09804
	(0.03001)	(0.02081)	**(0.00954)**	(0.00956)	(0.04773)	(0.02450)	(0.01214)
9	0.20391	0.11755	**0.09787**	0.10573	0.13488	0.24605	0.10261
	(0.02863)	(0.02756)	(0.01104)	**(0.00994)**	(0.04037)	(0.03780)	**(0.00916)**
10	0.20395	0.10625	0.09638	0.10233	0.12756	0.23490	0.09619
	(0.03225)	(0.02081)	(0.01042)	**(0.01009)**	(0.02889)	(0.02595)	(0.01046)

Note: The best result for each testing set is marked in bold, and the values in parentheses represent the standard deviation.

TABLE 6.5

Means and Standard Deviations of the Spread for 10 Sets of Testing Scenarios

Set No.	MOGA	NPGA2	NSGA2	SPEA2	PAES	MICROGA	JG
1	0.74152	0.74734	0.69743	**0.66582**	0.76670	0.75060	0.68446
	(0.10338)	(0.08215)	(0.06515)	(0.06933)	(0.08157)	**(0.06024)**	(0.07525)
2	0.73228	0.73442	0.68371	**0.68223**	0.75569	0.74729	0.69770
	(0.11134)	(0.09683)	**(0.05346)**	(0.07555)	(0.08871)	(0.08584)	(0.07438)
3	0.78992	0.76592	0.73156	**0.69784**	0.77206	0.76033	0.72092
	(0.08887)	(0.08058)	(0.07368)	**(0.05715)**	(0.06768)	(0.07240)	(0.06642)
4	0.73193	0.73335	0.68279	**0.65844**	0.74382	0.75308	0.68758
	(0.08227)	(0.10570)	(0.06545)	(0.07002)	(0.09125)	**(0.06503)**	(0.08574)
5	0.74873	0.73854	0.70099	0.68153	0.72993	0.75895	**0.66971**
	(0.09790)	(0.08932)	(0.06268)	(0.06495)	(0.08674)	(0.07685)	(0.09441)
6	0.73376	0.73185	**0.67238**	0.67747	0.74052	0.73717	0.69793
	(0.12482)	(0.09975)	**(0.06595)**	(0.09353)	(0.07645)	(0.06721)	(0.07733)
7	0.76183	0.78587	0.71150	0.70257	0.78087	0.77232	**0.70157**
	(0.06721)	(0.09185)	**(0.06167)**	(0.07376)	(0.07168)	(0.06634)	(0.06873)
8	0.78949	0.74535	0.69586	**0.68563**	0.77069	0.74163	0.71773
	(0.08135)	(0.09058)	(0.07698)	**(0.06053)**	(0.07041)	(0.07680)	(0.06799)
9	0.75589	0.73199	0.69594	**0.68536**	0.74917	0.77237	0.69144
	(0.09985)	(0.08328)	(0.06500)	(0.06547)	(0.07845)	**(0.05840)**	(0.06584)
10	0.72555	0.74602	0.70190	**0.67218**	0.75851	0.75435	0.69335
	(0.09500)	(0.09902)	**(0.05988)**	(0.07854)	(0.08901)	(0.06254)	(0.06924)

Note: The best result for each testing set is marked in bold, and the values in parentheses represent the standard deviation.

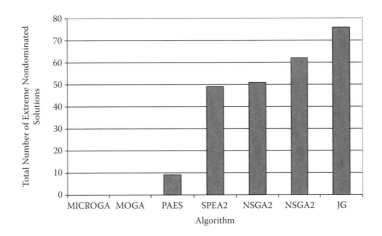

FIGURE 6.5
Total number of extreme nondominated solutions for each MOEA obtained from 10 sets of testing scenarios.

The following summarizes the results:

1. As compared with the MOGA, NPGA2, and MICROGA, the JG was more favorable for all 10 sets.
2. As compared with the NSGA2, the JG was more favorable for sets 1, 5, 6, 7, and 10. However, it was inconclusive for the remaining sets.
3. As compared with the SPEA2, the JG was more favorable for sets 4 and 6–8. Nevertheless, it was inconclusive for the remaining sets.
4. As compared with the PAES, the JG was more favorable for sets 3 and 10. However, it was inconclusive for the remaining sets.

From these outcomes, the JG scored 41 favorable marks and 19 inconclusive marks for all 10 testing sets. As a consequence, it was able to acquire better sets of nondominated solutions than other MOEAs with good convergence and diversity performance. Sample nondominated solution sets searched by various MOEAs for the 10 sets of testing scenarios are depicted in Figure 6.6 for reference.

In conclusion, based on different performance metrics, JG was statistically better than the other MOEAs as reflected by the ε-indicator. It performed better in convergence, and although it had a small effect in diversity, it obtained more extreme cases in the reference Pareto optimal set.

TABLE 6.6

Number of Occurrences in Three Different Cases for 10 Sets of Testing Scenarios

Different WLL Scenarios	Cases	MOGA	NPGA2	NSGA2	SPEA2	PAES	MICROGA
Set 1	Case I	**1,599**	1,306	**649**	360	593	**1,600**
	Case II	0	10	498	508	0	0
	Case III	1	284	453	**732**	**1,007**	0
Set 2	Case I	**1,599**	1,272	481	520	532	b
	Case II	0	24	431	342	0	0
	Case III	1	304	**688**	**738**	1,068	0
Set 3	Case I	**1,600**	1,272	482	473	**941**	**1,600**
	Case II	0	24	379	441	0	0
	Case III	0	304	**739**	**686**	659	0
Set 4	Case I	**1,587**	1,241	488	**645**	499	**1,600**
	Case II	0	15	378	466	1	0
	Case III	13	344	**734**	489	**1,100**	0
Set 5	Case I	**1,600**	1,361	**598**	386	529	**1,600**
	Case II	0	13	433	481	0	0
	Case III	0	226	569	**733**	**1,071**	0
Set 6	Case I	**1,598**	1,145	**592**	632	424	**1,600**
	Case II	0	56	454	463	0	0
	Case III	2	399	554	505	**1,176**	0
Set 7	Case I	**1,600**	1,353	**666**	616	555	**1,600**
	Case II	0	6	379	370	0	0
	Case III	0	241	555	614	**1,045**	0
Set 8	Case I	**1,599**	1,257	408	**638**	462	**1,600**
	Case II	0	17	580	550	0	0
	Case III	1	326	**612**	412	**1,138**	0
Set 9	Case I	**1,600**	1,285	388	469	465	**1,600**
	Case II	0	52	573	408	0	0
	Case III	0	263	**639**	**723**	**1,135**	0
Set 10	Case I	**1,600**	1,296	**693**	583	920	**1,600**
	Case II	0	10	377	304	0	0
	Case III	0	294	530	**713**	680	0

Note: The best result for each testing set is marked in bold.

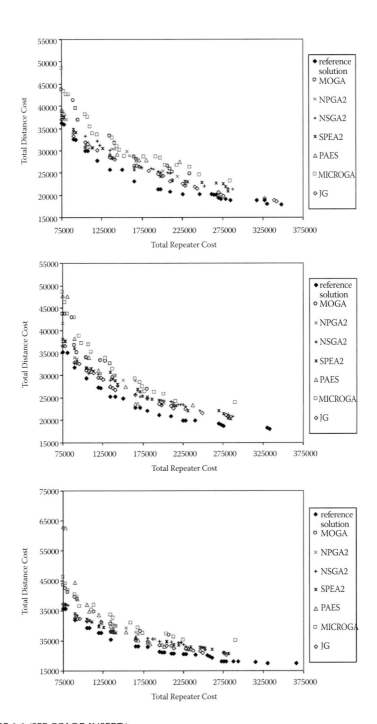

FIGURE 6.6 (SEE COLOR INSERT.)
Sample nondominated solution sets obtained from each MOEA for 10 sets of testing scenarios.

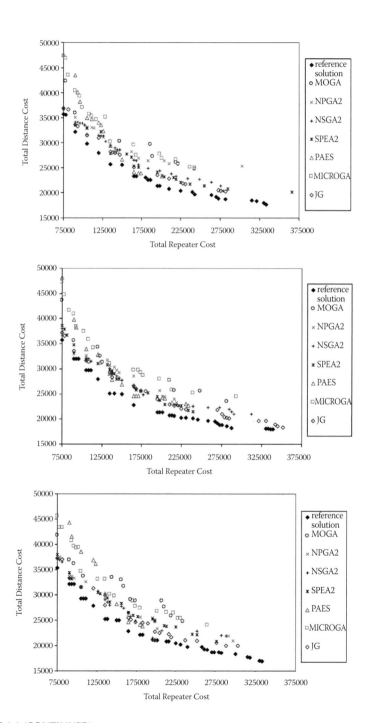

FIGURE 6.6 (CONTINUED)

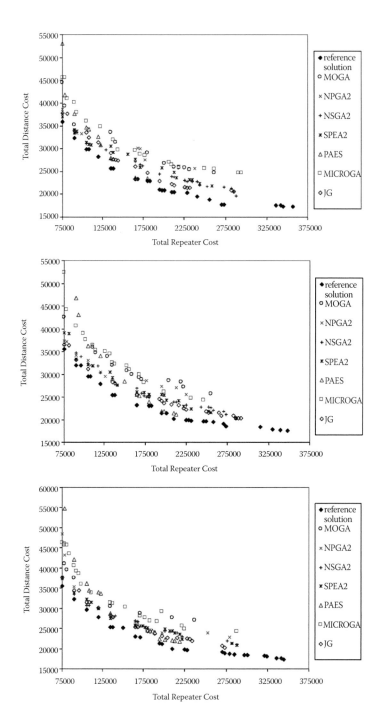

FIGURE 6.6 (CONTINUED)

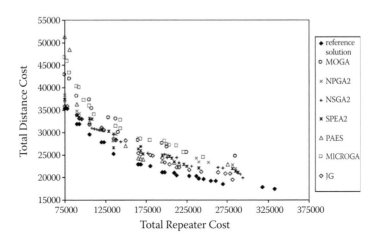

FIGURE 6.6 (CONTINUED)

References

1. Bae, G. S., Son, H. K., A study of 2.3 GHz propagation characteristic measured in Korea, in *Proceedings of International Symposium on Antennas and Propagation Society*, Columbus, OH, June 2003, 2:995–998.
2. Barbounakis, S., Stavroulakis, P., Gardiner, J. G., General aspects of digital technologies for wireless local loops, *International Journal of Communication Systems*, 13(3), 187–206, 2000.
3. Black, U., *Second Generation Mobile and Wireless Networks*, Upper Saddle River, NJ: Prentice Hall, 1999.
4. Chan, T. M., Man, K. F., Tang, K. S., Kwong, S., Multiobjective optimization of radio-to-fiber repeater placement using a jumping gene algorithm, in *Proceedings of IEEE International Conference on Industrial Technology*, Hong Kong, China, December 2005, 291–296.
5. Cox, D. C., Wireless loops: What are they? *International Journal of Wireless Information Networks*, 3(3), 125–138, 1996.
6. Gagnaire, M., *Broadband Local Loops for High-Speed Internet Access*, Boston: Artech House, 2003.
7. Garg, V. K., *Wireless Network Evolution: 2G to 3G*, Upper Saddle River, NJ: Prentice Hall, 2002.
8. Garg, V. K., Sneed, E. L., Digital wireless local loop system, *IEEE Communications Magazine*, 34(10), 112–115, 1996.
9. Jeong, D. G., Jeon, W. S., Current and future services using wireless local loop (WLL) systems, *International Journal of Communication Systems*, 13(3), 289–301, 2000.
10. Lee, W. C. Y., Spectrum and technology of a wireless local loop system, *IEEE Personal Communications Magazine*, 5(1), 49–54, 1998.
11. Lin, Y. B., The wireless local loop, *IEEE Potentials*, 16(3), 8–10, 1997.

12. Lin, Y. B., Chlamtac, I., *Wireless and Mobile Network Architectures*, New York: Wiley, 2001.
13. Man, K. F., Tang, K. S., Kwong, S., *Genetic Algorithms: Concepts and Designs*, Berlin: Springer, 1999.
14. Muller, N. J., *Mobile Telecommunications Factbook*, New York: McGraw-Hill, 1998.
15. Noerpel, R., WLL: Wireless local loop—Alternative technologies, in *Proceedings Eighth IEEE International Symposium on Personal, Indoor and Mobile Radio Communications*, Helsinki, Finland, September 1997, 2:619–623.
16. Noerpel, R., Lin, Y. B., Wireless local loop: Architecture, technologies and services, *IEEE Personal Communications*, 5(3), 74–80, 1998.
17. Park, S. K., Song, P. J., Bae, G. S., Joint optimization of radio repeater location and linking in WLL systems with 2.3 GHz frequency band, in *Proceedings IEEE International Conference on Communications*, Vancouver, Canada, June 1999, 3:1617–1621.
18. Stavroulakis, P., *Wireless Local Loops: Theory and Applications*, Chichester, UK: Wiley, 2001.
19. Walke, H., *Mobile Radio Networks: Networking, Protocols and Traffic Performance*, Chichester, UK: Wiley, 2002.
20. Webb, W., *Understanding Cellular Radio*, Boston: Artech House, 1998.
21. Webb, W., *Introduction to Wireless Local Loop: Broadband and Narrowband Systems*, Boston: Artech House, 2000.
22. Westerveld, R., Prasad, R., Rural communications in India using fixed cellular radio systems, *IEEE Communications Magazine*, 32(10), 70–77, 1994.
23. Xia, H. H., A simplified analytical model for predicting path loss in urban and suburban environments, *IEEE Transactions on Vehicular Technology*, 46(4), 1040–1046, 1997.

7

Resource Management in WCDMA

7.1 Introduction

Due to the recent tremendous demand for Internet and multimedia services, third-generation (3G) mobile communication systems are being widely adopted. 3G systems provide wireless Internet services, including information and Web browsing, mobile banking, mobile shopping, and multimedia services with voice, data, and video. Better system capacity and service quality are now available due to the high data transmission rate of the system owing to the wide bandwidth allocated [8,15,16,19].

The transmission speed of 3G systems varies with different situations. It is 144 kbps for high-speed movement (e.g., users in vehicles); 384 kbps for low-speed movement (e.g., pedestrians); and 2 Mbps for stationary status (e.g., users sitting in an office) [1,18,24], supporting both packet-switched and circuit-switched services.

Many different proposals have been suggested by different organizations [3,4,7,17,21], and they are summarized in Table 7.1. The four radio transmission technologies—cdma2000, WCDMA-FDD (wideband code division multiple access frequency division duplex) (UTRA-FDD); WCDMA-time division duplex (WCDMA-TDD) (UTRA-TDD), and Enhanced Data Rates for GSM Evolution (EDGE)—that could fulfill 3G system requirements were considered [9,13,14,25]. However, their radio parameters and the protocol stack with respect to the requirement for global roaming were incompatible [18].

In this chapter, the WCDMA system is the focus. For any wireless mobile communication system, the interests of users and operators are somehow contradictory. From the users' point of view, they hope to obtain good quality of service (QoS). The QoS reflects the signal quality of connecting users and can be represented in a number of ways, such as bit error rate (BER), received bit energy-to-noise density ratio E_b/N_0, and so on. In our study, E_b/N_0 was considered since it is more commonly used [10–12].

For different services, such as voice, data, or video services, a minimum value of E_b/N_0 (denoted γ) is to be specified for each type of service. If the E_b/N_0 value of a user is equal to or greater than that of the requested service,

TABLE 7.1

Different Radio Transmission Technologies Proposed by Various Organizations

Organization	Proposal of Radio Transmission Technology
European Telecommunications Standards Institute (ETSI) in Europe	WCDMA-frequency division duplex (WCDMA-FDD)
	WCDMA-time division duplex (WCDMA-TDD)
	Digital enhanced cordless telephone (DECT)
Association of Radio Industries and Businesses (ARIB) in Japan	WCDMA-FDD
Telecommunications Industry Association (TIA) in the United States	Universal Wireless Communications 136 (UWC-136)
	cdma2000
	Wireless multimedia and messaging services WCDMA (WIMS WCDMA)
T1P1 in the United States	WCDMA-North America (WCDMA-NA)
Telecommunications Technology Association (TTA) in Korea	WCDMA cdma2000
China Academy of Telecommunications Technologies (CATT) in China	Time division synchronous CDMA (TD-SCDMA)

the QoS of the user is then satisfied, and hence good service with guaranteed quality is available.

On the other hand, service operators want to optimize their revenues by maximizing the total number of users utilizing their networks (i.e., the total capacity). The maximum capacity of a network depends on the requested service and the E_b/N_0 value of each user. Consider an extreme case for which all users only use the voice service; a large capacity can be obtained as voice service demands fewer resources compared to video service. Nevertheless, this is unlikely to happen because users usually ask for different services.

In general, network users request voice, data, and video services simultaneously. If each user enjoys the requested service with $\frac{E_b}{N_0} \gg \gamma$, more resources are consumed, and the maximum capacity will decrease.

To fulfill the desires of both service operators and users better, resource management is indeed important from a system design point of view. The objective is to allocate the resources optimally so that the radio spectrum can be efficiently utilized; the QoS of each connecting user can be satisfied, and the network capacity is maximized [5,26–28].

The network maximum capacity depends on the E_b/N_0 value of each user, which in turn is related to the resources, including transmission power and transmission rate. As a result, resource management in transmission power and rate control is required.

It is known that if the assigned transmission power of a particular user is too high, it will lower the service quality of the others. A similar result applies to the user's transmission rate. If one's transmission rate is too high, one's QoS may be easily violated. Thus, by adjusting the transmission power and transmission rate of users, the E_b/N_0 value of a QoS-violating user can be improved, resulting in better QoS. In addition, having the E_b/N_0 value of a QoS-satisfying user close to γ can eventually improve the QoS of other users. Thus, it is desirable that each user adopt a sufficient transmission power and transmission rate so that the QoS is satisfied.

Assume that no users end their calls, and no idle users initiate a new call at a particular moment; resource management can reallocate the excessive resources consumed by QoS-satisfying users to QoS-violating users. Furthermore, this can also provide sufficient resources for users who connect in the future if the maximum capacity has not yet been reached.

In this study, a centralized resource management in direct sequence wideband code division multiple access (DS-WCDMA) systems was considered. Its implementation steps were as follows: At the beginning of a time slot, the resource management center received the data from the database, including the measured link gain, the requested media type, and the identification number of the assigned base station of each connecting user. With these data, the center performed an optimization technique to determine the optimized transmission power and transmission rate of each connecting user within the maximum allowable computation time. Finally, each base station received the corresponding values for the transmission power and transmission rate of its connecting mobile phones from the center and instructed each phone to apply the values by sending a command through its communication channel.

Note that the total transmission power and the total transmission rate were competing with each other. Therefore, the problem was a multiobjective optimization problem.

7.2 Mathematical Formulation

In this section, an optimization problem is formulated so that resources, including transmission powers and transmission rates, are optimally allocated to all connecting users [2]. The objectives are to minimize the total transmission power and maximize the total transmission rate of all users simultaneously. However, optimization of these two objectives may easily lead to solutions with a large number of users violating the QoS. Thus, an alternative method is sought. The problem is reformulated by introducing

TABLE 7.2

List of Notations

Notations	Meaning
N	A set of connecting users $\{1,2,3,\cdots,|N|\}$
W	Total spread-spectrum bandwidth
η	Background noise power
g_{bi}	Link gain between the mobile user i and the base station b $\forall i \in N$
p_i	Transmission power of user i $\forall i \in N$
r_i	Transmission rate of user i $\forall i \in N$
P_i^{min}	Minimum allowed transmission power of user i $\forall i \in N$
P_i^{max}	Maximum allowed transmission power of user i $\forall i \in N$
R_i^{min}	Minimum allowed transmission rate of user i $\forall i \in N$
R_i^{max}	Maximum allowed transmission rate of user i $\forall i \in N$
$\left(\dfrac{E_b}{N_o}\right)_i$	Received bit energy-to-noise density ratio of user i $\forall i \in N$
γ_i	Required minimum value of the received bit energy-to-noise density ratio of user i $\forall i \in N$
T	A constant considered to be larger than the total transmission rate of $|N|$ users
λ_p	The fixed cost per unit power in watts
λ_r	The fixed cost per unit rate in kilobits per second
λ_x	The fixed cost per user who violates the QoS requirement
x_i	$x_i = 1$ if user i violates the QoS requirement and 0 otherwise, $\forall i \in N$

Source: Data from Chan, T. M., Man, K. F., Tang, K. S., Kwong, S., A jumping gene algorithm for multiobjective resource management in wideband CDMA systems, *The Computer Journal*, 48(6), 749–768, 2005.

an extra objective to minimize the total number of QoS violating users. Its purpose is to direct the search to solutions with minimal total power and maximal total rate while keeping the total number of QoS-violating users at a minimum. The notations used are given in Table 7.2, and the three objectives are as follows:

$$\text{Total transmission power: } f_1 = \lambda_p \cdot \sum_{i \in N} p_i \tag{7.1}$$

$$\text{Total transmission rate: } f_2 = \lambda_r \cdot \left(T - \sum_{i \in N} r_i \right) \tag{7.2}$$

$$\text{Total number of violating users: } f_3 = \lambda_x \cdot \sum_{i \in N} x_i \tag{7.3}$$

subject to the following constraints:

1. The transmission power of each user must lie in the range between the minimum and maximum allowed power: $P_i^{min} \le p_i \le P_i^{max}, \forall i \in N$.
2. The transmission rate of each user must lie in the range between the minimum and maximum allowed rate: $R_i^{min} \le r_i \le R_i^{max}, \forall i \in N$.
3. Fulfill the QoS requirement of each user (each user's received bit energy-to-noise density ratio must be greater than or equal to a required minimum value $\gamma_i, \forall i \in N$):

$$\left(\frac{E_b}{N_o}\right)_i = \frac{g_{bi} \cdot p_i / r_i}{\left(\displaystyle\sum_{\substack{j \in N \\ j \neq i}} g_{bj} \cdot p_j + \eta\right) \Big/ W} \ge \gamma_i, \forall i \in N .$$

It should be noted that the objective function of Equation (7.2) is modified so that the task simply minimizes all three objective functions f_1, f_2, f_3 as given in Equations (7.1)–(7.3).

7.3 Chromosome Representation

The design of the chromosome is depicted in Figure 7.1; the transmission power and transmission rate of each connecting user are encoded in floating point and considered as the genes of chromosomes. Since the transmission power and the transmission rate of each user are independent, the same values are possible as found in a chromosome.

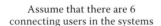

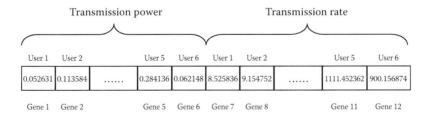

FIGURE 7.1
Chromosome-encoding method. (From Chan, T. M., Man, K. F., Tang, K. S., Kwong, S., A jumping gene algorithm for multiobjective resource management in wideband CDMA systems, *The Computer Journal*, 48(6), 749–768, 2005. With permission from Oxford University Press.)

7.4 Initial Population

As the first step in a genetic algorithm (GA), a population of initial chromosomes (solutions) is generated. The following discussion presents how the power and the rate values of each user in the chromosomes are generated.

7.4.1 Power Generation

In the power generation scheme, the transmission power of a user is to be generated within two new specified power ranges with regard to the link gains instead of the range between minimum and maximum allowable power. This is because users with larger link gains should utilize lower power to reduce the multiple access interference of other users. This can thus improve their signal-to-noise ratios and reduce the chance of violating their QoS constraints.

In contrast, those with smaller link gains should utilize higher powers for acquiring the QoS constraints they demand. However, a purely random value within the specified minimum and maximum allowable power failed to manage this, and it is common to have a large number of QoS-violating users.

The proposed transmission power generation rule was designed as follows: If the link gain of a user is greater than or equal to a threshold value α, the power value is generated randomly between $[p_{i\min}, p_{border}]$, where $p_{i\min}$ and p_{border} are the initial minimum power and the borderline power, respectively. Otherwise, it is generated randomly between $[p_{border}, p_{i\max}]$, where $p_{i\max}$ is the initial maximum power.

A suitable set of values of α, $p_{i\min}$, p_{border}, and $p_{i\max}$, which produced better results (i.e., more solutions have zero QoS-violating users), could be selected after testing and comparing their different combinations. In our simulations, these values were set as $\alpha = 100$, $p_{i\min} = 0.001$, $p_{border} = 0.005$, and $p_{i\max} = 0.01$.

7.4.2 Rate Generation

Since the transmission rate of a user does not affect the signal-to-noise ratio of the others, it is simply randomly generated between the guaranteed and the peak rates of the requested media type.

Note that even though the power and rate generation method given may not be the best, it is able to improve search performance.

7.5 Jumping Gene Transposition

There are four possible types of genes (for transmission power, voice user transmission rate, data user transmission rate, and video user transmission rate) in a chromosome. Thus, four corresponding types of transposons can be designed but can only jump within the specified range through the cut-and-paste or copy-and-paste operation.

For example, as shown in Figure 7.1, assuming that there are six connecting users in the system, the ranges of the transmission power gene, voice user transmission rate gene, data user transmission rate gene, and video user transmission rate gene are between gene 1 and gene 6, between gene 7 and gene 8, between gene 9 and gene 10, and between gene 11 and gene 12, respectively. A chromosome will be infeasible if genes jump to any position outside their corresponding ranges. Therefore, the jumping positions must be restricted.

7.6 Mutation

In this problem, the conventional mutation method did not perform well, as it cannot effectively improve the QoS of those QoS-violating users appearing in a chromosome. First, this is because the operational rate is usually small; hence only a small number of QoS-violating users may have their transmission power and transmission rates altered. On the other hand, if a high mutation rate is set, the GA changes to possess a random searching behavior and no longer is properly guided. Second, even if the genes have undergone mutation, it cannot be guaranteed that improvement always happens if the gene is simply replaced by some random values.

To overcome these problems and improve searching performance, a new mutation operation is introduced. Three different cases are considered, and the decision of how the transmission power and transmission rate are to be changed is based on the value of $\left(\frac{E_b}{N_o}\right)$ (see Section 7.2), that is, whether the QoS constraint of the users is satisfied or violated.

Case A: $\left(\dfrac{E_b}{N_o}\right)_i > \gamma_i$, for $i \in N$ **(i.e., the QoS Constraint Is Satisfied)**

The transmission power or transmission rate genes of user i will be mutated if the corresponding random number is less than the mutation rate pm. Let the current gene values for transmission power and transmission rate of user i be p_i and r_i, respectively; a new transmission value is applied by generating a random number within the range of $\left[P_i^{\min}, p_i\right]$, and the transmission rate is replaced by a new random value obtained within the range of $\left[r_i, R_i^{\max}\right]$.

This design tries to decrease the transmission power and increase the transmission rate of a user, respectively. This is because power reduction can decrease the multiple access interference of other users and hence increase their $\left(\frac{E_b}{N_o}\right)$. Consequently, the chance of QoS-violating users becoming QoS-satisfying users is increased. Also, an increase of rate value can let user i enjoy faster service without affecting the QoS of other users.

Case B: $\left(\dfrac{E_b}{N_o}\right)_i < \gamma_i$, for $i \in N$ (i.e., the QoS Constraint Is Violated)

Both the power and the rate genes of user i will be mutated. The changes of the power and the rate values are based on the following rules: The new power value is generated randomly between $\left[p_i, P_i^{max}\right]$, while the new transmission rate is generated randomly between $\left[R_i^{min}, r_i\right]$.

The purpose of these changes is to increase the transmission power and to decrease the transmission rate, respectively. The value of $\left(\frac{E_b}{N_o}\right)_i$ of user i is then improved; hence the chance that the QoS of user i is satisfied is increased. However, it should be noted that increasing power may cause violation of the QoS constraint of the other users since multiple access interference is also increased.

Case C: $\left(\dfrac{E_b}{N_o}\right)_i = \gamma_i$, for $i \in N$ (i.e., the QoS Constraint Is Just Satisfied)

Since this is the ideal situation, the transmission power and transmission rate of user i will remain unchanged, even if mutation occurs. In fact, any modification will cause certain adverse effects as follows:

1. An increase of power: The QoS constraints of other users may be violated since the increase of a user's power will increase the multiple access interference of the others.
2. A reduction of power: The QoS constraint of user i may be violated.
3. An increase of rate: The QoS constraint of user i may be violated.
4. A reduction of rate: This contradicts one of the goals (i.e., the maximization of total rate).

In summary, the new mutation is operated as follows:

1. After crossover, evaluate $\left(\frac{E_b}{N_o}\right)$ of all users in each of the new chromosomes.
2. Set the chromosome counter $k = 1$.
3. Set the user counter $i = 1$.
4. If user i belongs to case A, complete the corresponding actions and go to step 7.
5. If user i belongs to case B, complete the corresponding actions and go to step 7.
6. Otherwise, keep the power and rate genes of user i unchanged.

7. $i = i + 1$. If $i \leq N$, go to step 4.
8. $k = k + 1$. If $k \leq N_p$, go to step 3.
9. End.

An example of this operation is given in Figure 7.2. Assume that 25 users are connecting to the system; there exist $2 \times 25 = 50$ genes in the chromosome as shown. Suppose that only genes 1, 2, and 50 are going to be mutated; it is deduced that user 1 (voice user), user 2 (voice user), and user 25 (video user) belong to cases A, B, and C, respectively. After mutation, the value of gene 1 (the power value of user 1) is decreased. The values of gene 2 (the power value of user 2) and gene 27 (the rate value of user 2) are increased and reduced, respectively. However, gene 50 (the rate value of user 25) will not be mutated and remains unchanged as the value of $\left(\frac{E_b}{N_o}\right)$ belongs to case C.

7.7 Ranking Rule

The standard Pareto ranking (see Section 2.2.1 in Chapter 2) may not be suitable for this optimization problem. This is because the third objective f_3 (i.e., the total number of QoS-violating users) is more crucial than the other two objectives [see Equations (7.1)–(7.3) for the definition of the objectives].

Let us consider two chromosomes I_1 and I_2 with objective values f_1, f_2, f_3 and f_1', f_2', f_3', respectively [f_1, f_2, f_3 are defined in Equations (7.1)–(7.3)]. If $f_3 < f_3'$, I_1 should be preferable to I_2 regardless of the values of f_1, f_2, f_1', and f_2'. Therefore, the ranking rule is slightly modified as follows:

I_1 is preferable to I_2 if and only if

$$(f_3 < f_3') \text{ or } (\forall m = 1,2,3 \ f_m \leq f_m' \text{ and } \exists n = 1,2 \text{ s.t. } f_n < f_n')$$

The main benefit of this modified ranking rule is that it can direct the search toward solutions with fewer QoS-violating users or even zero such users, while it can identify multiple trade-off solutions for different total numbers of QoS-violating users.

7.8 Results and Discussion

The parameters of WCDMA systems and user services are assumed and are listed in Tables 7.3 and 7.4, respectively. The system model contains 9 hexagonal cells, as shown in Figure 7.3. Each cell has a base station located at the

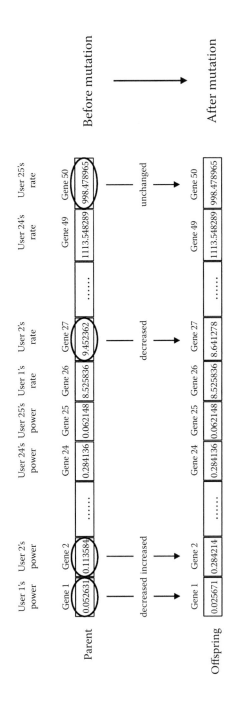

FIGURE 7.2
Mutation. (From Chan, T. M., Man, K. F., Tang, K. S., Kwong, S., A jumping gene algorithm for multiobjective resource management in wideband CDMA systems, *The Computer Journal*, 48(6), 749–768, 2005. With permission from Oxford University Press.)

TABLE 7.3

Parameters of WCDMA Systems

Parameter	Value
The total number of users $|N|$	Scenario a: 25
	Scenario b: 50
Total spread-spectrum bandwidth W	5 MHz
Background noise power η	−174 dBm/Hz
Cell radius	1 km

Source: Data from Chan, T. M., Man, K. F., Tang, K. S., Kwong, S., A jumping gene algorithm for multiobjective resource management in wideband CDMA systems, *The Computer Journal*, 48(6), 749–768, 2005.

TABLE 7.4

Parameters of User Services

Parameter	Service Type Voice	Data	Video
Minimum allowed transmission power P^{min}	0 W	0 W	0 W
Maximum allowed transmission power P^{max}	1 W	1 W	1 W
Minimum allowed transmission rate R^{min}	8.4 kbps	50 kbps	844 kbps
Maximum allowed transmission rate R^{max}	9.6 kbps	144 kbps	1,125 kbps
Minimum required value of the received bit energy-to-noise density ratio	4.2 dB	3.7 dB	5 dB

Source: Data from Chan, T. M., Man, K. F., Tang, K. S., Kwong, S., A jumping gene algorithm for multiobjective resource management in wideband CDMA systems, *The Computer Journal*, 48(6), 749–768, 2005.

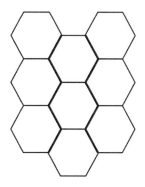

FIGURE 7.3
The WCDMA system model containing nine hexagonal cells. (From Chan, T. M., Man, K. F., Tang, K. S., Kwong, S., A jumping gene algorithm for multiobjective resource management in wideband CDMA systems, *The Computer Journal*, 48(6), 749–768, 2005. With permission from Oxford University Press.)

center, and the locations of all users are randomly distributed to all the cells. The distance between user i and base station b can be measured by

$$d_{bi} = \sqrt{(x_i - x_b)^2 + (y_i - y_b)}$$ (7.4)

where (x_i, y_i) and (x_b, y_b) are the x-y coordinates of user i and base station b, respectively.

The services provided for all users include voice, data, and video, and users' requests are randomly generated. The link gain g_{bi} between mobile user i and base station b is modeled as follows:

$$g_{bi} = \frac{s_{bi}}{d_{bi}^4}$$ (7.5)

where s_{bi} is the shadow fading factor, and d_{bi} is the distance between mobile user i and base station b.

The values of s_{bi} are generated through the log-normal distribution with an expected value of 0 dB and a standard deviation of 8 dB. For each of two scenarios, 25 users and 50 users, 10 testing sets of different distributions of user locations and different requested user services were considered in the simulation.

The JG is compared with multiobjective evolutionary algorithms (MOEAs), including the multiobjective genetic algorithm (MOGA), niched Pareto genetic algorithm 2 (NPGA2), nondominated sorting genetic algorithm 2 (NSGA2), strength Pareto evolutionary algorithm 2 (SPEA2), Pareto archived evolution strategy (PAES), and microgenetic algorithm (MICROGA), and their parameters are given in Table 7.5. Apart from these MOEAs, a modified GA, which only relies on the mutation operation proposed in Shayestch, Menhaj, and Nobary [20], is also included. They reported that its performance was much better than a common GA and close to the optimum detector; at the same time, it requires less computational complexity.

To realize the performance of the sets of nondominated solutions found by each MOEA, the two performance metrics (Deb and Jain convergence metric and spread, mentioned in Sections 5.2 and 5.3 of Chapter 5, respectively) were adopted. The means and standard deviations of the two performance metrics were obtained based on 50 simulation runs performed for each MOEA in each of 10 testing sets for the two different scenarios (Tables 7.6–7.9).

Since the true Pareto-optimal set is unknown for each case, a set of reference solutions was found to approximate the true Pareto-optimal set, as explained in Section 5.2. This reference set was obtained by incorporating the MOEAs together with some arbitrarily sufficient large number of generations or iterations. (It was set as 200 generations for MOGA, NPGA2, NSGA2, SPEA2, and JG; 10,000 iterations for PAES; and 2,500 iterations for MICROGA for both scenarios.)

TABLE 7.5

Parameters of MOEAs

Parameter	Value/Type
Population size	
MOGA, NPGA2, NSGA2, SPEA2, JG	50
PAES	1
MICROGA	4
Maximum generations or iterations	
MOGA, NPGA2, NSGA2, SPEA2, JG	100
PAES	5,000
MICROGA	1,250
Crossover type	Uniform crossover
Crossover rate	0.8
Mutation rate	Scenario a (25 users): 0.02
	Scenario b (50 users): 0.01
Other Settings for JG	
Jumping rate	0.02
Number of transposons	4
Length of transposon	1
Other Settings for SPEA2 or PAES	
Archive size	100
Depth	4
Other Settings for MICROGA	
Size of external memory	100
Size of population memory	80
Percentage of nonreplaceable memory	0.25
Replacement cycle	Every 25 iterations
Number of subdivisions of the adaptive grid	25
Number of iterations to achieve nominal convergence	4

Source: Data from Chan, T. M., Man, K. F., Tang, K. S., Kwong, S., A jumping gene algorithm for multiobjective resource management in wideband CDMA systems, *The Computer Journal*, 48(6), 749–768, 2005.

Note that the results of the PAES, MICROGA, and modified GA are not listed in Tables 7.7 to 7.12 because the QoS of some users was violated in some simulation runs of the two scenarios even in the final nondominated solutions. Therefore, they were disqualified.

7.8.1 Mean and Standard Deviation of Deb and Jain Convergence Metric for Evaluating Convergence

The means and standard deviations of the Deb and Jain convergence metric for the two scenarios are presented in Tables 7.6 and 7.7. For scenario a with 25 users, the smallest means (i.e., the better convergence) were obtained by

TABLE 7.6

Means and Standard Deviations of the Deb and Jain
Convergence Metric for Scenario a: 25 Users

Set	MOGA	NPGA2	NSGA2	SPEA2	JG
1	1.46457	0.87476	0.87716	1.01084	**0.85851**
	(0.34465)	(0.34772)	**(0.24376)**	(0.37437)	(0.28257)
2	5.54971	**2.96351**	3.11623	4.06308	3.21363
	(1.42756)	**(0.65139)**	(0.80334)	(1.51124)	(1.12461)
3	2.97277	**1.51712**	1.58659	1.91971	1.54879
	(1.53381)	(0.56122)	**(0.32315)**	(1.23594)	(0.37745)
4	2.17939	1.52314	1.39023	1.61352	**1.38370**
	(0.71769)	(0.63025)	**(0.32363)**	(0.48567)	(0.57063)
5	9.28464	**6.10951**	7.87668	9.39389	7.26696
	(4.93257)	(4.62266)	(3.91845)	(6.50109)	**(2.71774)**
6	33.91906	15.82890	27.61811	**15.03545**	30.76600
	(33.89666)	(28.87598)	(18.89024)	**(13.25995)**	(22.91944)
7	13.60818	9.03703	4.06747	8.45967	**3.61446**
	(7.78791)	(9.39618)	(1.55185)	(6.63293)	**(1.30963)**
8	12.45420	6.84593	6.47069	8.01524	**5.89585**
	(8.76255)	(5.07673)	(1.72789)	(5.07927)	**(1.69749)**
9	7.66898	3.33059	**2.54958**	4.73264	2.65271
	(5.90743)	(1.93979)	**(0.49800)**	(2.80844)	(0.72296)
10	3.48460	2.21659	**2.01528**	2.57183	2.02774
	(1.18591)	(1.16675)	(0.61998)	(0.93743)	**(0.52989)**

Note: The best result for each testing set is marked in bold, and the values in parentheses represent the standard deviation.

the JG in four sets (sets 1, 4, 7, and 8), while the NPGA2, NSGA2, and SPEA2 were the best in three sets (sets 2, 3, and 5), two sets (sets 9 and 10), and one set (set 6), respectively.

Regarding scenario b with 50 users, the JG was the winner in seven sets (sets 1–6 and 10). In contrast, the NPGA2 and NSGA2 won in two sets (sets 7 and 9) and one set (set 8), respectively. Moreover, by using the JG, better performance in the scenario with a larger number of total connecting users (i.e., scenario b with 50 users) was found.

7.8.2 Mean and Standard Deviation of Spread for Evaluating Diversity

The means and standard deviations of the spread for the two scenarios are shown in Tables 7.8 and 7.9. Although the JG did not acquire better diversity in any set for scenario a with 25 users, it was the winner of six sets (sets 1, 3, 4, 6, 7, and 9) for scenario b with 50 users.

TABLE 7.7

Means and Standard Deviations of the Deb and Jain Convergence Metric for Scenario b: 50 Users

Set	MOGA	NPGA2	NSGA2	SPEA2	JG
1	13.46179	8.39987	6.37275	10.43100	**6.06237**
	(5.71459)	(5.53798)	**(1.459121)**	(4.741348)	(1.89943)
2	59.60346	19.26178	26.96596	29.33367	**19.17224**
	(65.98679)	(9.28943)	(14.59247)	(14.247461)	**(7.71705)**
3	72.61046	50.28666	33.55322	51.05154	**29.49879**
	(42.08842)	(33.06023)	**(8.70762)**	(21.898426)	(9.25324)
4	303.3512	153.1214	73.5405	129.8568	**55.6418**
	(199.1566)	(153.0053)	(37.7913)	(101.1470)	**(30.6113)**
5	263.9020	150.9318	141.2099	165.5834	**126.0499**
	(103.2448)	(104.1002)	(44.1734)	(127.0334)	**(38.6520)**
6	41.96852	18.05840	11.92954	38.94970	**8.52393**
	(34.81254)	(21.10841)	(4.93901)	(36.85691)	**(3.46312)**
7	12.45827	**9.40626**	11.49188	10.84423	11.17905
	(1.63119)	(2.31925)	(2.20902)	(2.57163)	(1.99769)
8	15.10310	11.17514	**8.75691**	11.81171	9.06114
	(6.57848)	(6.66243)	**(1.21806)**	(5.25236)	(1.42337)
9	12.35986	**8.03292**	8.42076	10.52409	8.42551
	(1.81740)	(1.88473)	**(1.80773)**	(5.03765)	(2.43144)
10	18.02363	8.84941	8.21734	10.98513	**7.57032**
	(8.17962)	(4.32090)	**(1.81956)**	(7.37767)	(2.21656)

Note: The best result for each testing set is marked in bold, and the values in parentheses represent the standard deviation.

As with the performance for convergence, the JG acquired better diversity performance in the scenario (i.e., scenario b) when more connecting users were encountered.

7.8.3 Diversity Evaluation Using Extreme Nondominated Solution Generation

Figures 7.4 and 7.5 depict the total number of extreme nondominated solutions found by various MOEAs in 50 simulation runs for all 10 sets of each of scenarios a and b, respectively. The scaling factors [see the definition given in Equations (5.11) and (5.12)] chosen for scenarios a and b were 0.01 and 0.02, respectively.

The reason to have a larger scaling factor for scenario b is that the distance between the nondominated fronts of the MOEAs and the reference front is larger. If 0.01 is assigned, the total number of extreme solutions found by each MOEA may be zero, and the results could not be compared.

TABLE 7.8

Means and Standard Deviations of the Spread for Scenario a: 25 Users

Set	MOGA	NPGA2	NSGA2	SPEA2	JG
1	0.95034	**0.94873**	0.94898	0.96654	0.94987
	(0.06094)	**(0.05389)**	(0.07279)	(0.11871)	(0.06413)
2	**0.96014**	0.99236	0.98297	0.98892	0.97441
	(0.05018)	(0.03982)	**(0.03309)**	(0.06712)	(0.05575)
3	0.94045	0.95223	**0.93142**	0.95442	0.94474
	(0.07583)	(0.07347)	(0.07374)	(0.10045)	**(0.05788)**
4	0.94631	**0.94435**	0.96509	0.95215	0.94769
	(0.05344)	(0.05736)	(0.05694)	(0.07179)	(0.06225)
5	**0.95138**	0.97284	0.97092	0.95252	0.98036
	(0.03794)	(0.03293)	(0.03196)	(0.04456)	**(0.02898)**
6	**0.95047**	0.96472	0.95385	0.99543	0.96520
	(0.04935)	(0.10818)	(0.05415)	(0.09183)	**(0.04194)**
7	**0.94276**	0.94743	0.96991	0.95669	0.94472
	(0.04559)	**(0.03828)**	(0.04925)	(0.05746)	(0.05503)
8	**0.94984**	0.96476	0.96755	0.96515	0.95749
	(0.05085)	**(0.04630)**	(0.05599)	(0.06331)	(0.05258)
9	0.94979	0.94490	0.96567	**0.93241**	0.94621
	(0.04538)	(0.05724)	(0.06228)	(0.06651)	(0.05164)
10	**0.94702**	0.97148	0.96736	0.96485	0.95994
	(0.04071)	**(0.03519)**	(0.04306)	(0.08309)	(0.06114)

Note: The best result for each testing set is marked in bold, and the values in parentheses represent the standard deviation.

From the figures, the results indicate that the total number of extreme non-dominated solutions for both scenarios was largest when the JG was used.

7.8.4 Statistical Test Using Binary ε-Indicator

Tables 7.10 and 7.11 show the corresponding statistical results of the binary ε-indicator in terms of the number of occurrences of three cases under the 10 testing sets for scenarios a and b, respectively. The three comparison cases are specified as follows:

Case I: The JG is better than the compared algorithm.

Case II: The JG is worse than the compared algorithm.

Case III: They are incomparable.

The results can be summarized as follows:

Scenario a: 25 users

1. As compared with the MOGA, the JG was more favorable, except for sets 5 and 6.

TABLE 7.9

Means and Standard Deviations of the Spread for Scenario b: 50 Users

Testing Set	MOGA	NPGA2	NSGA2	SPEA2	JG
1	0.96062	0.96917	0.96843	0.97096	**0.96003**
	(0.03261)	**(0.02618)**	(0.02822)	(0.02906)	(0.02876)
2	**0.95888**	0.96347	0.96531	0.96491	0.97416
	(0.03379)	(0.03510)	(0.03890)	(0.04128)	(0.062043)
3	0.95964	0.96828	0.97048	0.96120	**0.95817**
	(0.03512)	**(0.02732)**	(0.03533)	(0.04391)	(0.04916)
4	0.95982	0.96202	0.96421	0.96377	**0.94788**
	(0.03662)	(0.04374)	(0.05022)	(0.04380)	(0.05851)
5	**0.95541**	0.96625	0.96334	0.95894	0.96291
	(0.03169)	(0.03527)	**(0.02834)**	(0.04291)	(0.03669)
6	0.95817	0.96369	0.96771	0.95566	**0.95168**
	(0.04278)	**(0.03486)**	(0.03583)	(0.03866)	(0.05337)
7	0.97214	0.98094	0.97076	0.97222	**0.97002**
	(0.02222)	(0.02363)	(0.02729)	(0.02588)	(0.03060)
8	**0.95921**	0.96735	0.97458	0.96891	0.96095
	(0.03750)	(0.02611)	(0.02456)	**(0.02443)**	(0.03080)
9	0.96348	0.96340	0.96429	0.96015	**0.95274**
	(0.03106)	(0.03829)	(0.03538)	(0.03922)	(0.04628)
10	**0.95928**	0.96656	0.97195	0.95944	0.96126
	(0.03114)	**(0.02769)**	(0.03023)	(0.03351)	(0.03570)

Note: The best result for each testing set is marked in bold, and the values in parentheses represent the standard deviation.

2. As compared with the NPGA2, the JG was more favorable, except for sets 5 and 6.

3. As compared with the NSGA2, the JG was more favorable for all 10 sets.

4. As compared with the SPEA2, the JG was more favorable, except for sets 5, 6, and 10.

Scenario b: 50 users

1. As compared with the MOGA, the JG was more favorable, except for set 5.

2. As compared with the NPGA2, the JG was more favorable, except for sets 5–7.

3. As compared with the NSGA2, the JG was more favorable, except for set 1.

4. As compared with the SPEA2, the JG was more favorable, except for set 5.

TABLE 7.10

Statistical Results of Binary ε-Indicator in Terms of the Number of Occurrences in Three Comparison Cases for Scenario a: 25 Users

Testing Set	Cases	MOGA	NPGA2	NSGA2	SPEA2
1	Case I	**2,006**	1,317	1,144	1,376
	Case II	194	495	641	524
	Case III	300	688	715	600
2	Case I	**1,881**	1,013	1,243	982
	Case II	327	705	676	682
	Case III	292	782	581	836
3	Case I	**2,147**	1,148	1,403	1,183
	Case II	97	529	586	656
	Case III	256	823	511	661
4	Case I	**1,807**	1,305	1,184	1,210
	Case II	273	534	663	635
	Case III	420	661	653	655
5	Case I	643	364	**1,280**	896
	Case II	690	**1,108**	541	637
	Case III	**1,167**	1,028	679	967
6	Case I	643	398	**1,159**	452
	Case II	872	1,007	715	992
	Case III	985	**1,095**	626	1,056
7	Case I	**2,411**	2,036	1,378	1,911
	Case II	28	213	594	261
	Case III	61	251	528	328
8	Case I	**1,641**	1,056	1,506	1,300
	Case II	312	747	507	500
	Case III	547	697	487	700
9	Case I	**1,974**	1,134	1,100	1,616
	Case II	71	515	645	339
	Case III	455	851	755	545
10	Case I	**1,325**	898	913	637
	Case II	586	770	763	878
	Case III	589	832	824	**985**

Source: Data from Chan, T. M., Man, K. F., Tang, K. S., Kwong, S., A jumping gene algorithm for multiobjective resource management in wideband CDMA systems, *The Computer Journal*, 48(6), 749–768, 2005.

Note: The best result for each testing set is marked in bold.

TABLE 7.11

Statistical Results of Binary ε-Indicator in Terms of the Number of Occurrences in Three Comparison Cases for Scenario b: 50 Users

Testing Set	Cases	MOGA	NPGA2	NSGA2	SPEA2
1	Case I	**2,316**	**1,334**	815	**2,089**
	Case II	102	416	734	233
	Case III	82	750	**951**	178
2	Case I	**1,667**	**859**	**1,127**	**1,099**
	Case II	423	811	680	703
	Case III	410	830	693	698
3	Case I	**2,108**	**1,575**	**1,102**	**1,750**
	Case II	105	389	659	414
	Case III	287	536	739	336
4	Case I	**2,333**	**1,662**	**1,243**	**1,801**
	Case II	70	451	626	278
	Case III	97	387	631	421
5	Case I	719	354	**1,142**	363
	Case II	488	**1,089**	676	1,033
	Case III	**1,293**	1,057	682	**1,104**
6	Case I	**1,642**	478	**1,175**	**1,277**
	Case II	285	902	671	527
	Case III	573	**1,120**	654	696
7	Case I	**1,343**	616	**1,230**	**945**
	Case II	428	854	559	698
	Case III	729	**1,030**	711	857
8	Case I	**2,060**	**1,239**	**974**	**1,398**
	Case II	130	646	761	441
	Case III	310	615	765	661
9	Case I	**1,965**	**1,022**	**1,049**	**1,513**
	Case II	226	719	667	462
	Case III	309	759	784	525
10	Case I	**2,289**	**1,094**	**993**	**1,419**
	Case II	62	534	666	420
	Case III	149	872	841	661

Source: Data from Chan, T. M., Man, K. F., Tang, K. S., Kwong, S., A jumping gene algorithm for multiobjective resource management in wideband CDMA systems, *The Computer Journal*, 48(6), 749–768, 2005.

Note: The best result for each testing set is marked in bold.

TABLE 7.12

Means and Standard Deviations of Running Times for Four Cases (ms)

	Case 1	Case 2	Case 3	Case 4
Average	283.64	314.06	545.2	591.68
Standard deviation	17.81	45.16	36.70	64.39

Source: Data from Chan, T. M., Man, K. F., Tang, K. S., Kwong, S., A jumping gene algorithm for multiobjective resource management in wideband CDMA systems, *The Computer Journal*, 48(6), 749–768, 2005.

In conclusion, the JG scored 33 favorable marks, 6 inconclusive marks, and only 1 unfavorable mark for scenario a with 25 users. Regarding scenario b with 50 users, the JG scored 34 favorable marks, 5 inconclusive marks, and only 1 unfavorable mark. Therefore, it was demonstrated that the JG had a powerful and effective search ability to seek better sets of nondominated solutions with better performance in convergence and diversity.

For illustration, some sampled nondominated solution sets obtained by different MOEAs for the 10 sets of each of scenarios a and b are shown in Figures 7.6 and 7.7, respectively.

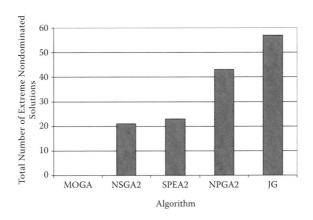

FIGURE 7.4
Total number of extreme nondominated solutions obtained by different MOGAs for scenario a: 25 users. (From Chan, T. M., Man, K. F., Tang, K. S., Kwong, S., A jumping gene algorithm for multiobjective resource management in wideband CDMA systems, *The Computer Journal*, 48(6), 749–768, 2005. With permission from Oxford University Press.)

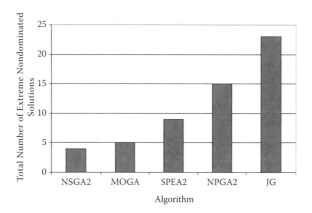

FIGURE 7.5

Total number of extreme nondominated solutions obtained by different MOGAs for scenario b: 50 users. (From Chan, T. M., Man, K. F., Tang, K. S., Kwong, S., A jumping gene algorithm for multiobjective resource management in wideband CDMA systems, *The Computer Journal*, 48(6), 749–768, 2005. With permission from Oxford University Press.)

7.9 Discussion of Real-Time Implementation

The integrated scheme comprised a GA, and the fast closed-loop power control was proposed [11] to achieve a possible real-time implementation. This power control adopted by the Interim Standard 95 (IS-95) [22] was suggested in 3G proposals [6,23]. Furthermore, it was indicated that adjusting the control period of the proposed scheme to 0.1 s was affordable for current microprocessors [11].

Based on the similar control scheme, the optimization scheme using JG is also workable in real-time environments with rapid changes. The factors affecting the computational complexity of the JG are (1) the total number of connecting users in the network, which increases the chromosome length, and (2) the transposition operations, which increase the computational load. To test how fast the JG can work with variations of these factors, the statistical results of running times were collected by carrying out 50 simulation runs using a Pentium IV 1.3-GHz computer for the following four different configurations:

1. Total number of connecting users is 25, and no JG transposition is implemented (i.e., transposition rate is zero);
2. Total number of connecting users is 25, and JG transposition is implemented;

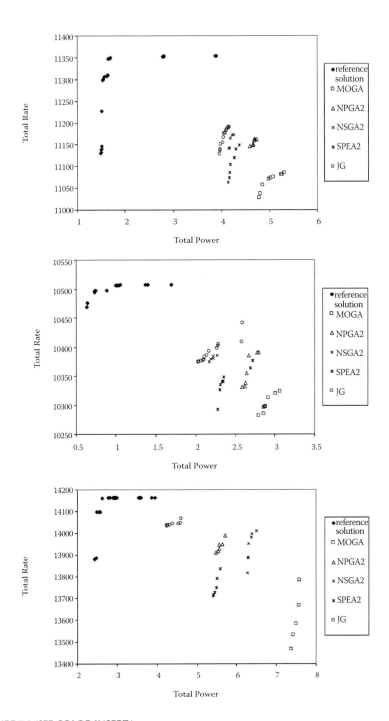

FIGURE 7.6 (SEE COLOR INSERT.)
Sampled nondominated solution sets obtained by different MOGAs for scenario a: 25 users.

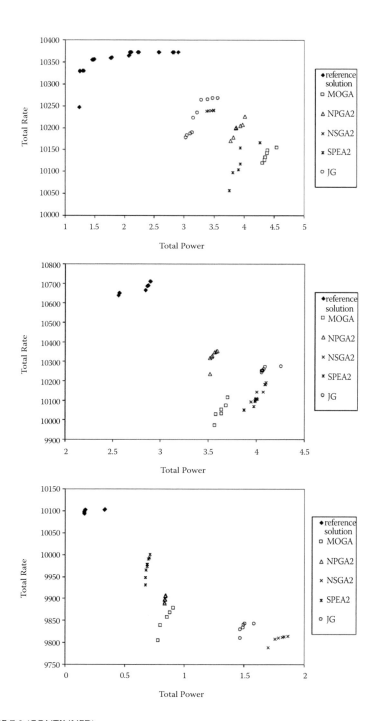

FIGURE 7.6 (CONTINUED)

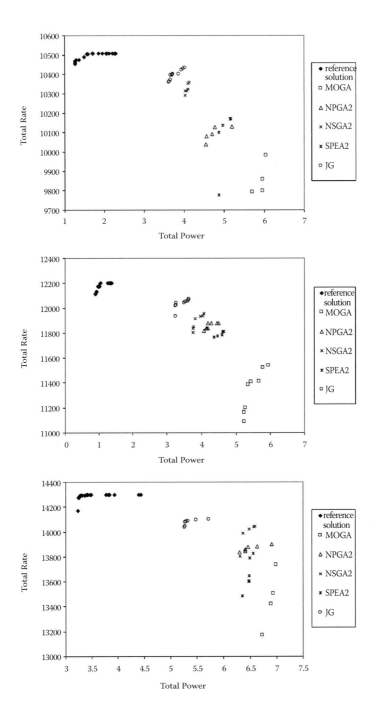

FIGURE 7.6 (CONTINUED)

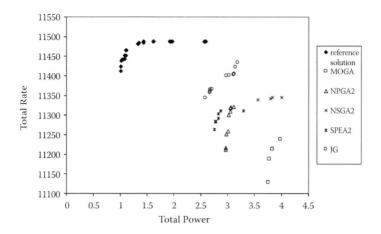

FIGURE 7.6 (CONTINUED)

3. Total number of connecting users is 50, and no JG transposition is implemented; and

4. Total number of connecting users is 50, and JG transposition is implemented.

Note that the parameters of the JG for this test were the same as in Table 7.5. The mean and standard deviation of running times are shown in Table 7.12. The average differences between configurations 1 and 3 and configurations 2 and 4 were 261.56 ms and 277.62 ms, respectively, indicating that the runtime increased with the total number of connecting users, as expected.

In addition, the average differences between configurations 1 and 2 and configurations 3 and 4 were 30.42 ms and 46.48 ms, respectively, indicating about a 10% increment of the runtime when the JG transposition was implemented. The runtime only slightly increased as the transposition rate was indeed small. The small standard deviations also implied that the runtime of the JG was quite stable.

Undoubtedly, it is hoped that the JG can obtain better performance by performing more generations when a faster processor is adopted. In real implementation, the JG can be forced to give optimal power and rate values for each connecting user when the maximum allowed time is reached, no matter how the total number of connecting users varies.

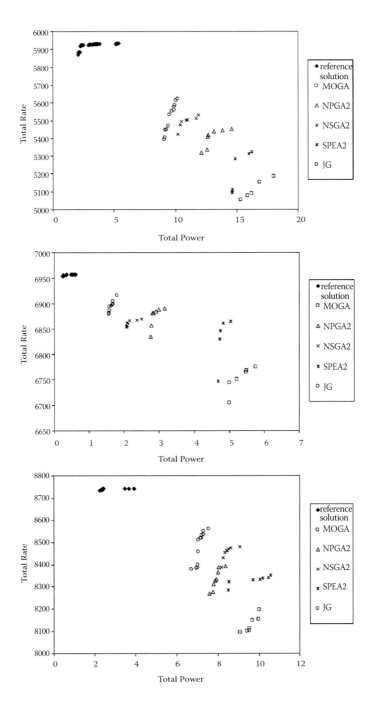

FIGURE 7.7 (SEE COLOR INSERT.)
Sampled nondominated solution sets obtained by different MOGAs for scenario b: 50 users.

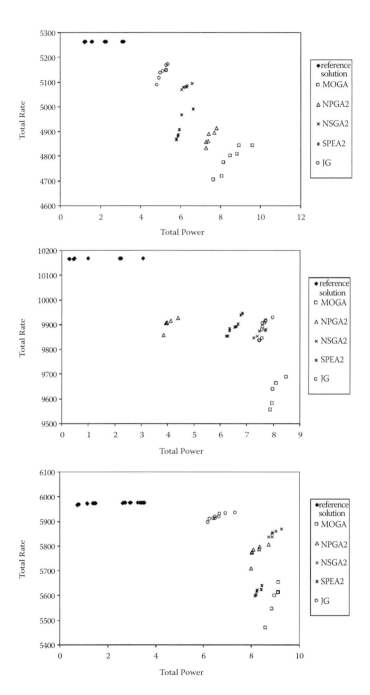

FIGURE 7.7 (CONTINUED)

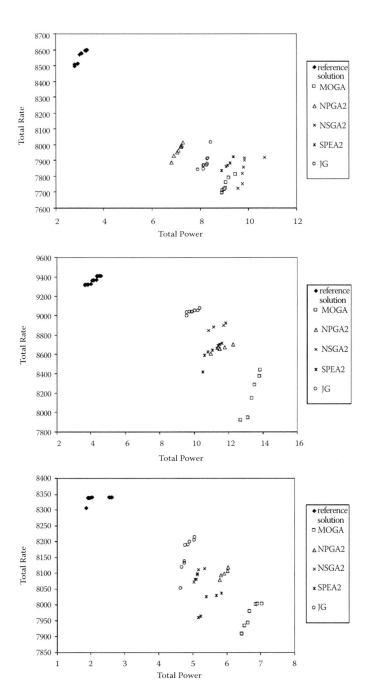

FIGURE 7.7 (CONTINUED)

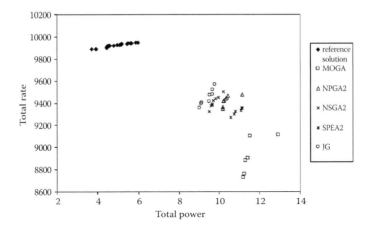

FIGURE 7.7 (CONTINUED)

References

1. Andreadis, A., Giambene, G., *Protocols for High-efficiency Wireless Networks*, Boston: Kluwer Academic, 2003.
2. Chan, T. M., Man, K. F., Tang, K. S., Kwong, S., A jumping gene algorithm for multiobjective resource management in wideband CDMA systems, *The Computer Journal*, 48(6), 749–768, 2005.
3. Chaudhury, P., Mohr, W., Onoe, S., The 3GPP proposal for IMT-2000, *IEEE Communications Magazine*, 37(2), 72–81, 1999.
4. Dixit, S., Prasad, R. (Eds.), *Wireless IP and Building the Mobile Internet*, Boston: Artech House, 2003.
5. Dixit, S., Guo, Y., Antoniou, Z., Resource management and quality of service in third generation wireless networks, *IEEE Communications Magazine*, 39(2), 125–133, 2001.
6. ETSI/UTRA, *The ETSI UMTS Terrestrial Radio Access (UTRA) ITU-R RTT Candidate Submission*, Sophia-Antipolis Cedex, France: European Telecommunications Standards Institute, June 1998.
7. Huber, J. F., Weiler, D., Brand, H., UMTS, the mobile multimedia vision for IMT 2000: A focus on standardization, *IEEE Communications Magazine*, 38(9), 129–136, 2000.
8. Lin, Y. B., Chlamtac, I., *Wireless and Mobile Network Architectures*, New York: Wiley, 2001.
9. Lloyd-Evans, R., *QoS in Integrated 3G Networks*, Boston: Artech House, 2002.
10. Moustafa, M., Habib, I., Naghshineh, M., Wireless resource management using genetic algorithm for mobiles equilibrium, in *Proceedings of Sixth IEEE Symposium on Computers and Communications*, Hammamet, Tunisia, June 2001, 586–591.

11. Moustafa, M., Habib, I., Naghshineh, M., GAME based radio resource management in wideband CDMA networks, in *Proceedings of IEEE Global Telecommunications Conference*, San Antonio, TX, December 2001, 6:3618–3622.
12. Moustafa, M., Naghshineh, M., Genetic algorithm for mobiles equilibrium, in *Proceedings of 21st Century Military Communications Conference*, Los Angeles, CA, October 2000, 1:70–74.
13. Nicopolitidis, P., Obaidat, M. S., Papadimitriou, G. I., Pomportsis, A. S., *Wireless Networks*, Chichester, UK: Wiley, 2003.
14. Nilsson, M., Third-generation radio access standards, *Ericsson Review*, 76(3), 110–121, 1999.
15. Nilsson, T., Toward third-generation mobile multimedia communication, *Ericsson Review*, 76(3), 122–131, 1999.
16. Oliphant, M. W., The mobile phone meets the Internet, *IEEE Spectrum*, 36(8), 20–28, 1999.
17. Pandya, R., *Mobile and Personal Communication Services and Systems*, New York: IEEE Press, 2000.
18. Prasad, R., Mohr, W., Konhauser, W. (Eds.), *Third Generation Mobile Communication Systems*, Boston: Artech House, 2000.
19. Sarikaya, B., Packet mode in wireless networks: Overview of transition to third generation, *IEEE Communications Magazine*, 38(9), 164–172, 2000.
20. Shayesteh, M. G., Menhaj, M. B., Nobary, B. G., A modified genetic algorithm for multiuser detection in DS/CDMA systems, *IEICE Transactions on Communications*, E86-B(8), 2377–2388, 2003.
21. Sheikh, A. U. H., *Wireless Communications: Theory and Techniques*, Boston: Kluwer Academic, 2004.
22. TIA/EIA/IS-95, *Mobile Station—Base Station Compatibility Standard for Dual-Mode Wideband Spread Spectrum Cellular Systems*, Arlington, VA: Telecommunications Industry Association, 1995.
23. TIA/TR45.5.4, *The cdma2000 ITU-R RTT Candidate Submission (0.18)*, Arlington, VA: Telecommunications Industry Association, 1998.
24. Wang, J. Z., *Broadband Wireless Communications: 3G, 4G and Wireless LAN*, Boston: Kluwer Academic, 2001.
25. Wesolowski, K., *Mobile Communication Systems*, Chichester, UK: Wiley, 2002.
26. Zander, J., Radio resource management in future wireless networks: Requirements and limitations, *IEEE Communications Magazine*, 35(8), 30–36, 1997.
27. Zander, J., Kim, S. L., *Radio Resource Management for Wireless Networks*, Boston: Artech House, 2001.
28. Zhang, J. H., Huai, J. P., Xiao, R. Y., Li, B., Resource management in the next-generation DS-CDMA cellular networks, *IEEE Wireless Communications*, 11(4), 52–58, 2004.

8

Base Station Placement in WLANs

8.1 Introduction

Nowadays, wireless technology has penetrated most local-area networks (LANs). Wireless local-area networks (WLANs) have become common and popular. They are one of the most deployed wireless networks in the world and are widely installed in venues such as homes, schools, enterprises, hospitals, airports, retail outlets, manufacturing, corporate environments, and hot spot environments [1,6,12,13].

There are many reasons why WLANs have become so popular [3–5,13,14], and some are listed next:

User mobility: WLANs offer free movement to users who do not want to or even cannot access the network at fixed locations (e.g., classrooms, libraries, offices, shopping centers, factories, etc.). Thus, they can bring convenience to users.

Flexibility and scalability: WLANs can be flexibly adjusted to different network topologies to fulfill specific applications and installations. Unlike wired LANs, the topology of WLANs can be easily modified if necessary. Furthermore, only authorizations, but not cables, are required for adding users to WLANs. If WLANs are to be expanded, only additional access points and extension points are required.

Wiring difficulty: Floors in some buildings (e.g., factories, places containing hazardous materials) or locations outdoors are not suitable for installing wires for wired LANs. In contrast, there are many fewer constraints for WLANs.

Low overall cost of operation: The wiring cost in a wired LAN depends on the total number of stations that it can support and the distances between the stations and the hub. In contrast, there is no wiring cost for WLANs regardless of the total number and location of stations. Furthermore, their installation and maintenance costs are cheaper compared to those incurred by traditional additions, deletions, and changes of wired LANs.

Without cable fault: In wired LANs, moisture can cause erosion of metallic conductors and then breakage of cables. However, this problem does not occur in WLANs. As a result, WLANs help reduce network failures and avoid the costs of cable replacement.

In WLANs, the nodes send and receive data through one of three wireless transmission media: infrared, radio frequency (RF), and microwaves [7,13]. Although infrared-based WLANs and microwave-based WLANs have their own advantages in terms of cost or throughput, RF-based WLANs are still more practical and the most popular choice for indoor use as the RF can propagate through solid obstacles such as walls and floors [9].

There are two modes of operation for WLANs: infrastructure mode and ad hoc mode [8,10]. In the infrastructure mode, stations in the same cell communicate with each other via an access point that is used to transmit, receive, and buffer the data between the WLAN and the wired network infrastructure. Stations from different cells can also talk with each other through various access points. Therefore, the access points act like the base stations in a mobile cellular system.

When operating in ad hoc mode, each station can directly communicate with others through the access points (i.e., peer-to-peer communication is adopted), but within the same valid transmission ranges.

This chapter provides information to facilitate the installation of a WLAN in an integrated circuit (IC) factory. In this study, an infrastructure-based WLAN was considered for establishment in the factory; therefore, the locations of its access points (base stations) had to be correctly determined to provide adequate quality radio coverage for terminals.

Since there is a trade-off between the signal quality of terminals and the number of allowable base stations, this placement problem is better formulated as a multiobjective optimization problem.

8.2 Path Loss Model

To evaluate the total path loss and satisfy coverage performance, the log-distance path loss model is employed [15–19]. The average total path loss is expressed as a function of distance with respect to the path loss exponent n:

$$\overline{PL}(d) \propto \left(\frac{d}{d_0} \right)^n \tag{8.1}$$

or

$$\overline{PL}(d) = \overline{PL}(d_0) + 10n \log\left(\frac{d}{d_0}\right) \tag{8.2}$$

where d_0 is the reference distance; d is the distance between the base station and terminal; n is the path loss exponent indicating the rate at which the path loss increases with distance; $\overline{PL}(d_0)$ is the reference path loss due to free-space propagation from the base station to a 1-m reference distance that is equal to $10n_0 \log\left(\frac{4\pi d_0}{\lambda}\right)$; n_0 is the reference path loss exponent; and λ is the wavelength of the frequency used.

Since the reference distance d_0 is chosen as 1 m and the reference path loss exponent n_0 is equal to 2 for the free-space environment [16, 19],

$$\overline{PL}(d_0) = 20 \log\left(\frac{4\pi}{\lambda}\right) \tag{8.3}$$

For the obstructed environment in factories, the path loss exponent n is selected as 2 [16,19]. Moreover, because physical obstructions lie directly between the base station and terminal, an additional term, the penetration loss, has to be added to Equation (8.2) to account for the total average path loss. Then, $\overline{PL}(d)$ is finally modified as

$$\overline{PL}(d) = 20 \log\left(\frac{4\pi}{\lambda}\right) + 20 \log(d) + \sum_{i=1}^{J} N_i L_i$$

$$= 20 \log\left(\frac{4\pi d}{\lambda}\right) + \sum_{i=1}^{J} N_i L_i \tag{8.4}$$

where J is the total number of different types of obstructing objects; i is the type of obstructing objects; N_i is the number of obstructing objects with type i; and L_i is the penetration loss caused by an obstructing object of type i.

8.3 Mathematical Formulation

In this multiobjective base station placement problem, the mathematical model used in Tang, Man, and Kwong [19] was applied; the notations are listed in Table 8.1.

TABLE 8.1

List of Notations

Notation	Meaning
O	The set of coordinates inside the floor plan of the IC factory.
M	The set of the maximum allowable base stations.
B	The set of installed base stations.
T	The set of terminals and $|T|$ is the cardinality of T.
T_k	The set of terminals allocated to the installed base station k, $k \in B$.
b_j	The jth bit value in the control genes $j \in M$.
s	The maximum allowed power loss threshold.
pl_i	By assigning terminal i to an installed base station with the minimum average total path loss, this minimum average total path loss is equal to pl_i, $i \in T$.
o_k	The coordinates of the installed base station k, $k \in B$.
a_i	a_i is equal to 1 if pl_i is greater than s and 0 otherwise, $i \in T$.
c_{ik}	c_{ik} is equal to 1 if terminal i is assigned to the installed base station k and 0 otherwise, $i \in T$, $k \in B$.

Source: Data from Chan, T. M., Man, K. F., Tang, K. S., Kwong, S., A jumping-genes paradigm for optimizing factory WLAN network, *IEEE Transactions on Industrial Informatics*, 3(1), 33–43, 2007.

There are four objectives to be minimized:

1. The number of terminals with their path loss greater than the maximum allowed path loss threshold f_1:

$$f_1 = \sum_{i \in T} a_i \qquad (8.5)$$

2. The required number of base stations f_2:

$$f_2 = \sum_{j \in M} b_j \qquad (8.6)$$

3. The mean of the path loss predictions of the terminals f_3:

$$f_3 = \frac{1}{|T|} \sum_{i \in T} pl_i \qquad (8.7)$$

4. The mean of the maximum path loss predictions of the terminals in the set of allocated terminals of each base station f_4:

$$f_4 = \frac{1}{f_2} \sum_{k \in B} \max_{i \in R_k} (pl_i) \qquad (8.8)$$

subject to the following three constraints:

a. Each terminal i must be allocated to only one installed base station k: $\sum\limits_{k \in B} c_{ik} = 1, \ \forall i \in T$.

b. Any installed base station k must be located inside the floor plan of the IC factory: $o_k \in O, \ \forall k \in B$.

c. Only one base station is to be installed in a location.

8.4 Chromosome Representation

A hierarchical chromosome encoding scheme [11] is employed so that different total numbers of base stations to be installed were possible for different solutions. Similar to the design in Section 6.4, the hierarchical chromosome consists of two types of genes: control genes and parameter genes. The control genes are in binary form to control the activation of a base station, for which the x-y coordinates for installation are specified by a pair of parameter genes encoded in floating-point numbers.

Figure 8.1 depicts an example of a chromosome. Assuming that the maximum number of base stations that can be installed is 8, the hierarchical chromosome length is then 24 (8 control genes plus 16 parameter genes). If the ith control gene's value is 1, the location with the x coordinate in the $(i + 8)$th gene and the y coordinate in the $(i + 16)$th gene is chosen for installing a base station and is 0 otherwise, where $i = 1,2,\ldots,8$.

Referring to the example given in Figure 8.1, two base stations were installed at the locations with x-y coordinates (8.457125, 9.179531) and (31.583469, 19.247854) because their corresponding control genes were 1. However, no base station was to be installed at the x-y coordinates (20.371952, 17.671589) since the corresponding control gene is 0.

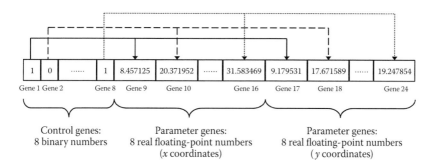

FIGURE 8.1
Encoding method for a hierarchical chromosome. (From Chan, T. M., Man, K. F., Tang, K. S., Kwong, S., A jumping-genes paradigm for optimizing factory WLAN network, *IEEE Transactions on Industrial Informatics*, 3(1), 33–43, 2007.)

8.5 Jumping Gene Transposition

There are two corresponding types of transposons, one for the control genes and one for the parameter genes. To ensure that the resultant chromosome after the jumping gene (JG) transposition is valid, the control transposon only presents and jumps within the range from the 1st to 8th genes, while the operations on the parameter transposon must be carried out between the 9th and 24th genes, as depicted in Figure 8.1.

8.6 Chromosome Repairing

After performing all the genetic operations, crossover, and mutation, the resultant offspring may violate the physical constraints given in Section 8.3. Since this chromosome now becomes invalid, it must be repaired and transformed into a valid one.

The repairing mechanism is given in Figure 8.2. In case (i), the x-y coordinates (31.583469, 3.989742) were assumed to be located outside the

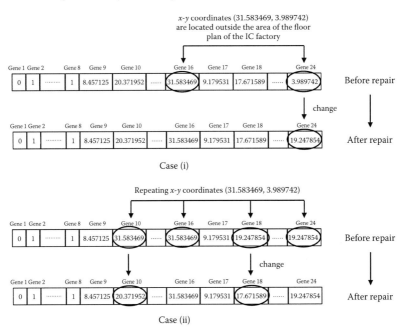

FIGURE 8.2
Repairing a chromosome. (From Chan, T. M., Man, K. F., Tang, K. S., Kwong, S., A jumping-genes paradigm for optimizing factory WLAN network, *IEEE Transactions on Industrial Informatics*, 3(1), 33–43, 2007.)

area of the floor plan of the IC factory. These invalid coordinates were then replaced by some random values; for example, the y coordinate was changed from 3.989742 to 19.247854, which was randomly generated, as shown in Figure 8.2. Similarly, in case (ii), repeated x-y coordinates, say (31.583469, 19.247854), were replaced by other randomly generated coordinates, say (20.371952, 17.671589), as only one base station can be installed in a particular location.

8.7 Results and Discussion

Without complicating the calculation and obscuring the essence of the proposed design approach, the WLAN design was based on a two-dimensional floor plan of the IC factory, as shown in Figure 8.3 [19]. The parameters of the WLAN and multiobjective evolutionary algorithms (MOEAs), including the MOGA (multiobjective genetic algorithm), NPGA2 (niched Pareto genetic algorithm 2), NSGA2 (nondominated sorting genetic algorithm 2), SPEA2 (strength Pareto evolutionary algorithm 2), PAES (Pareto archived evolution strategy), MICROGA (microgenetic algorithm), and JG, used in the study are shown in Tables 8.2 and 8.3, respectively. Two scenarios were defined; the

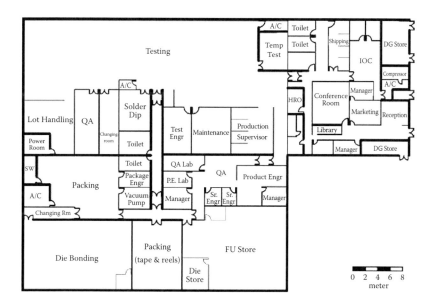

FIGURE 8.3
The floor plan of an IC factory. (From Tang, K. S., Man, K. F., Kwong, S., Wireless communication network design in IC factory, *IEEE Transactions on Industrial Electronics*, 48(2), 452–459, 2001.)

TABLE 8.2

Parameter Setting for a WLAN

Parameter	Value
Frequency	1.9 GHz
Maximum allowable base stations	8
Total terminals	238
Maximum allowed power loss threshold	Scenario a: 90 dB; scenario b: 80 dB
Penetration loss L_i	Type 1—thin partition: 2.0 dB
	Type 2—cement wall: 3.3 dB
	Type 3—thickened cement wall: 6.5 dB

Source: Data from Chan, T. M., Man, K. F., Tang, K. S., Kwong, S., A jumping-genes paradigm for optimizing factory WLAN network, *IEEE Transactions on Industrial Informatics*, 3(1), 33–43, 2007.

maximum allowed power loss threshold was set to be 90 dB for scenario a and 80 dB for scenario b. It was also assumed that every 3×3 m^2 section contained a possible terminal location.

To evaluate the performance of the nondominated solution sets found by each MOEA, the Deb and Jain convergence metric and the spread metric (see Sections 5.2 and 5.3 of Chapter 5, respectively) were considered. The means and standard deviations of the results were computed based on 50 simulation runs in each scenario.

Similar to the last two design examples given in Chapters 6 and 7, a set of reference solutions was found to approximate the true Pareto-optimal set for calculations of the performance metrics. This reference set was obtained by incorporating the MOEAs together with some sufficiently large number of generations or iterations (1,000 generations for MOGA, NPGA2, NSGA2, SPEA2, and JG; 100,000 iterations for PAES; and 25,000 iterations for MICROGA for both scenarios).

8.7.1 Mean and Standard Deviation of Deb and Jain Convergence Metric for Evaluating Convergence

The means and standard deviations of the Deb and Jain convergence metric for the two scenarios are shown in Table 8.4. It can be observed that the average metric values were the smallest (i.e., had the best convergence) when the JG was used for both scenarios.

8.7.2 Mean and Standard Deviation of Spread for Evaluating Diversity

Table 8.5 lists the means and standard deviations of the spread for the two scenarios. The JG was able to obtain the smallest mean values of the spread (i.e., had the best diversity) among the MOEAs for both scenarios.

TABLE 8.3

Parameter Settings for Different MOEAs

Parameter	Value/Type
Population Size	
MOGA, NPGA2, NSGA2, SPEA2, JG	100
PAES	1
MICROGA	4
Maximum Generations or Iterations	
MOGA, NPGA2, NSGA2, SPEA2, JG	500
PAES	50,000
MICROGA	12,500
Crossover type	Uniform crossover
Crossover rate	0.8
Mutation rate	0.04
Other Settings for JG	
Jumping rate	0.01
Number of transposons	3
Length of transposons	1
Other Settings for SPEA2 or PAES	
Archive size (SPEA2, PAES)	100
Depth (PAES)	4
Other Settings for MICROGA	
Size of external memory	100
Size of population memory	80
Percentage of nonreplaceable memory	0.25
Replacement cycle	Every 25 iterations
Number of subdivisions of the adaptive grid	25
Number of iterations to achieve nominal convergence	4

Source: Data from Chan, T. M., Man, K. F., Tang, K. S., Kwong, S., A jumping-genes paradigm for optimizing factory WLAN network, *IEEE Transactions on Industrial Informatics*, 3(1), 33–43, 2007.

8.7.3 Diversity Evaluation Using Extreme Nondominated Solution Generation

Figures 8.4 and 8.5 depict the total number of extreme nondominated solutions found by various MOEAs for scenarios a and b, respectively. In each scenario, nondominated solutions obtained by 50 simulation runs were considered, and the scaling factor chosen for both scenarios was 0.01. As shown in the figures, the total number of extreme nondominated solutions obtained by the JG were the largest for both scenarios.

TABLE 8.4

Means and Standard Deviations of the Deb and Jain Convergence Metric
for Scenarios a (90 dB) and b (80 dB)

Scenario	MOGA	NPGA2	NSGA2	SPEA2	PAES	MICROGA	JG
a: 90dB	0.10833	0.09613	0.08099	0.07837	0.13906	0.29506	**0.07679**
	(0.05951)	(0.04144)	(0.02254)	(0.04616)	(0.06530)	(0.03673)	**(0.01814)**
b: 80dB	0.12878	0.13489	0.10600	0.11018	0.14194	0.43459	**0.10277**
	(0.03607)	(0.04097)	**(0.02646)**	(0.03973)	(0.06876)	(0.05750)	(0.02709)

Note: The best result for each scenario is marked in bold, and the values in parentheses represent
the standard deviation.

TABLE 8.5

Means and Standard Deviations of the Spread for Scenarios a (90 dB) and b (80 dB)

Scenario	MOGA	NPGA2	NSGA2	SPEA2	PAES	MICROGA	JG
a: 90dB	0.98971	0.99518	0.97865	1.00420	1.00764	0.98532	**0.97854**
	(0.01251)	**(0.01001)**	(0.01751)	(0.02027)	(0.01095)	(0.01735)	(0.02118)
b: 80dB	0.96055	0.97978	0.94266	1.00016	1.03264	0.98283	**0.92105**
	(0.05362)	(0.04832)	(0.07118)	(0.06209)	**(0.03862)**	(0.04355)	(0.07829)

Note: The best result for each scenario is marked in bold, and the values in parentheses represent the standard deviation.

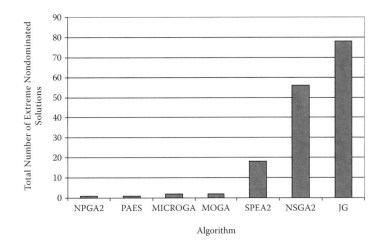

FIGURE 8.4
Total number of extreme nondominated solutions for scenario a: 90 dB. (From Chan, T. M.,
Man, K. F., Tang, K. S., Kwong, S., A jumping-genes paradigm for optimizing factory WLAN
network, *IEEE Transactions on Industrial Informatics*, 3(1), 33–43, 2007.)

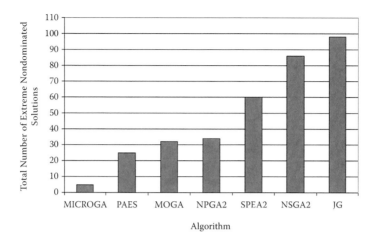

FIGURE 8.5

Total number of extreme nondominated solutions for scenario b: 80 dB. (From Tang, K. S., Kwong, S., Man, K. F., A jumping genes paradigm: Theory, verification and applications, *IEEE Circuits and Systems Magazine*, 8(4), 18–36, 2008.)

8.7.4 Statistical Test Using the Binary ε-Indicator

Table 8.6 shows the statistical results for the binary ε-indicator in terms of the number of occurrences of three comparison cases (same as those given in Section 7.8.4) for the two scenarios.

In conclusion, the JG was more favorable than other MOEAs for both scenarios. It obtained better sets of nondominated solutions with better convergence and diversity performance. Sample sets of nondominated solutions searched by different MOEAs for scenarios a and b are shown in Figure 8.6 to Figure 8.9 for reference.

TABLE 8.6

Statistical Results of Binary ε-Indicator in Terms of the Number of Occurrences in Three Different Cases for Scenarios a (90 dB) and b (80dB)

Scenarios	Case	MOGA	NPGA2	NSGA2	SPEA2	PAES	MICROGA
a: 90 dB	Case I	1,706	1,912	1,018	1,402	2,138	2,482
	Case II	137	82	642	366	16	0
	Case III	657	506	840	732	346	18
b: 80 dB	Case I	1,435	1,513	956	1,217	1,769	2,477
	Case II	143	131	860	297	0	14
	Case III	922	856	684	986	731	9

Source: Data from Chan, T. M., Man, K. F., Tang, K. S., Kwong, S., A jumping-genes paradigm for optimizing factory WLAN network, *IEEE Transactions on Industrial Informatics*, 3(1), 33–43, 2007.

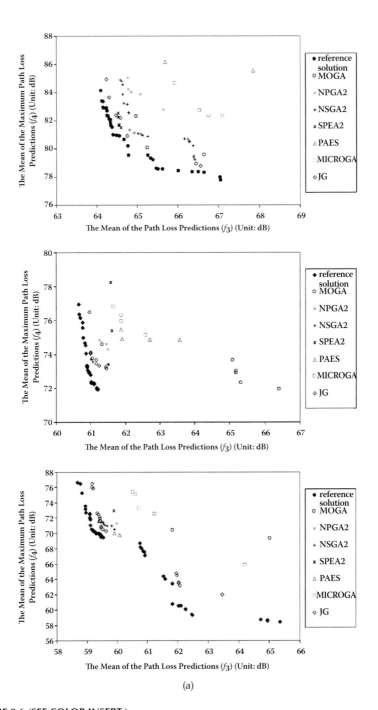

(a)

FIGURE 8.6 (SEE COLOR INSERT.)
Distribution of reference solutions and nondominated solutions obtained by MOGAs with different numbers of base stations for scenario a: 90 dB.

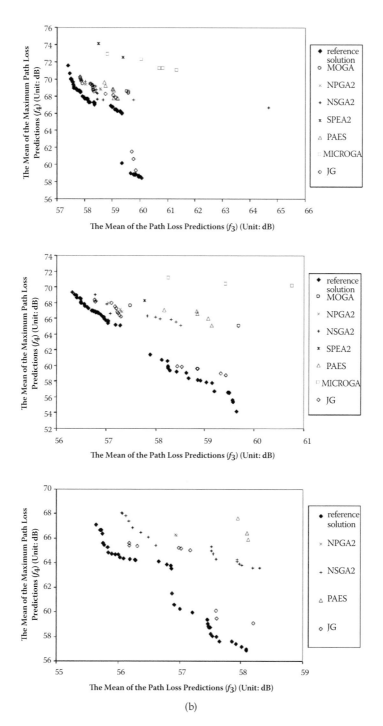

(b)

FIGURE 8.6 (CONTINUED)

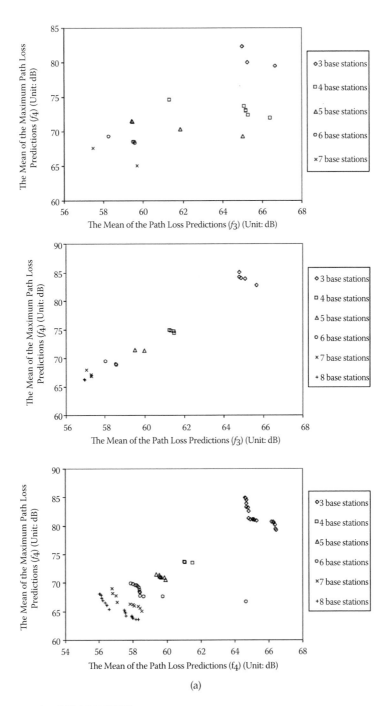

FIGURE 8.7 (SEE COLOR INSERT.)
Sample nondominated solution sets obtained by different MOGAs for scenario a: 90 dB.

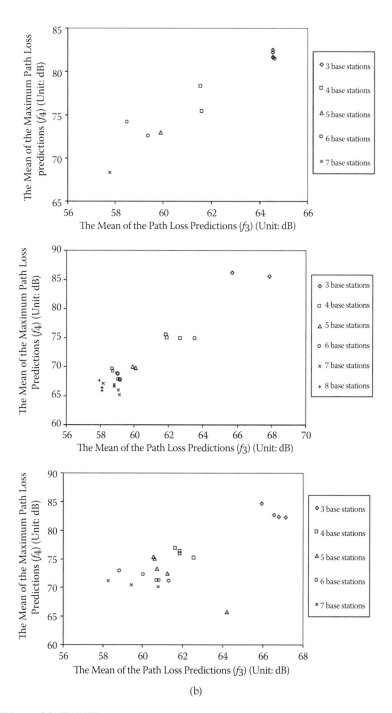

(b)

FIGURE 8.7 (CONTINUED)

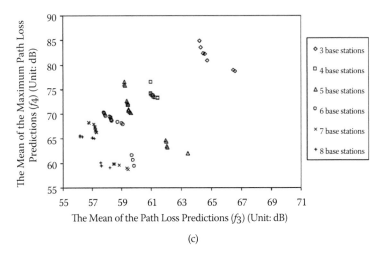

(c)

FIGURE 8.7 (CONTINUED)

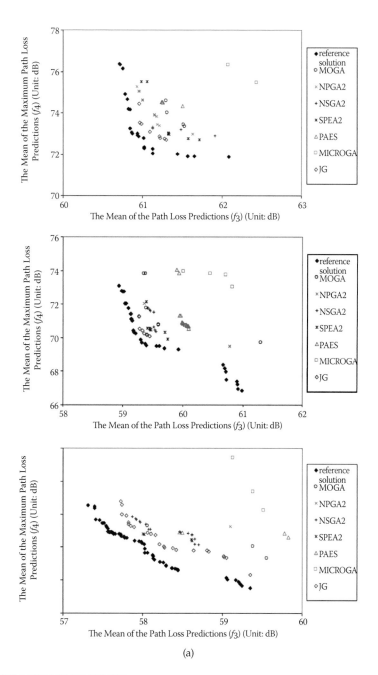

(a)

FIGURE 8.8 (SEE COLOR INSERT.)
Distribution of reference solutions and nondominated solutions obtained by MOGAs with different numbers of base stations for scenario a: 80 dB. (From Chan, T. M., Man, K. F., Tang, K. S., Kwong, S., A jumping-genes paradigm for optimizing factory WLAN network, *IEEE Transactions on Industrial Informatics*, 3(1), 33–43, 2007.)

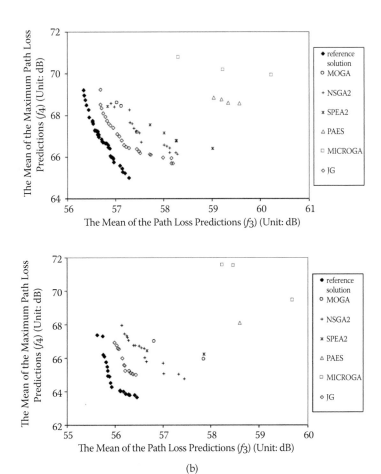

(b)

FIGURE 8.8 (CONTINUED)

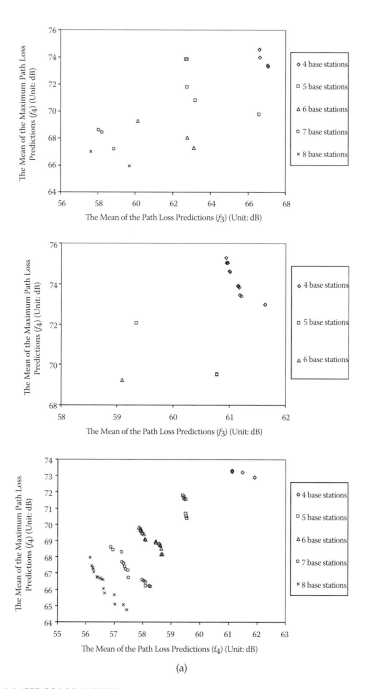

(a)

FIGURE 8.9 (SEE COLOR INSERT.)
Sample nondominated solution sets obtained by different MOGAs for scenario a: 80 dB. (From Chan, T. M., Man, K. F., Tang, K. S., Kwong, S., A jumping-genes paradigm for optimizing factory WLAN network, *IEEE Transactions on Industrial Informatics*, 3(1), 33–43, 2007.)

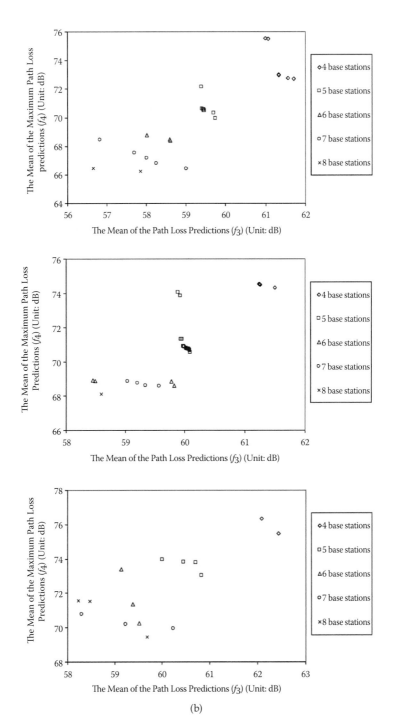

(b)

FIGURE 8.9 (CONTINUED)

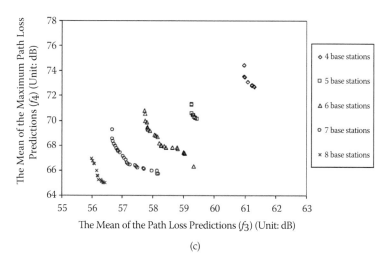

(c)

FIGURE 8.9 (CONTINUED)

References

1. Bing, B., Wireless local area networks: The new wireless revolution [Book review], *IEEE Communications Magazine*, 41(8), 8–10, 2003.
2. Chan, T. M., Man, K. F., Tang, K. S., Kwong, S., A jumping-genes paradigm for optimizing factory WLAN network, *IEEE Transactions on Industrial Informatics*, 3(1), 33–43, 2007.
3. Crow, B. P., Widjaja, I., Kim, L. G., Sakai, P. T., IEEE 802.11 wireless local area networks, *IEEE Communications Magazine*, 35(9), 116–126, 1997.
4. Gast, M. S., *802.11 Wireless Networks: The Definitive Guide*, Sebastopol, CA: O'Reilly, 2002.
5. Geier, J., *Wireless LANs*, Indianapolis, IN: Sams, 2002.
6. Gu, D., Zhang, J., QoS enhancement in IEEE802.11 wireless local area networks, *IEEE Communications Magazine*, 41(6), 120–124, 2003.
7. Held, G., *Deploying Wireless LANs: Concepts, Operation, and Utilization*, New York: McGraw-Hill, 2002.
8. Held, G., *Building a Wireless Network*, Boca Raton, FL: Auerbach, 2003.
9. Jordan, R., Abdallah, C. T., Wireless communications and networking: An overview, *IEEE Antennas and Propagation Magazine*, 44(1), 185–193, 2002.
10. Khan, J., Khwaja, A., *Building Secure Wireless Networks with 802.11*, Indianapolis, IN: Wiley, 2003.
11. Man, K. F., Tang, K. S., Kwong, S., *Genetic Algorithms: Concepts and Designs*, Berlin: Springer, 1999.
12. Mangold, S., Choi, S., Hiertz, G. R., Klein, O., Walke, B., Analysis of IEEE 802.11e for QoS support in wireless LANs, *IEEE Wireless Communications*, 10(6), 40–50, 2003.

13. Marincic, A., Milovanovic, D., Wireless local area networks, in *Proceedings Fourth International Conference on Telecommunications in Modern Satellite, Cable and Broadcasting Services*, Nis, Yugoslavia, October 1999, 1:291–299.
14. Prasad, A. R., Prasad, N. R., *802.11 WLANs and IP Networking: Security, QoS, and Mobility*, Boston: Artech House, 2005.
15. Rappaport, T. S., Characterization of UHF multipath radio channels in factory buildings, *IEEE Transactions on Antennas and Propagation*, 37(8), 1058–1069, 1989.
16. Rappaport, T. S. *Wireless Communications: Principles and Practice*, Upper Saddle River, NJ: Prentice Hall, 2002.
17. Saleh, A. A. M., Valenzuela, R. A., A statistical model for indoor multipath propagation, *IEEE Journal on Selected Areas in Communications*, 5(2), 128–137, 1987.
18. Tang, K. S., Kwong, S., Man, K. F., A jumping genes paradigm: Theory, verification and applications, *IEEE Circuits and Systems Magazine*, 8(4), 18–36, 2008.
19. Tang, K. S., Man, K. F., Kwong, S., Wireless communication network design in IC factory, *IEEE Transactions on Industrial Electronics*, 48(2), 452–459, 2001.

9

Conclusions

This book formulated all the necessary steps to ensure the use of the newly developed jumping gene (JG) algorithms. As JG is a new addition to the existing evolutionary algorithms, the fundamental materials (e.g., concept, theory, methods for verification, realistic as well as practical engineering applications) were sequentially presented.

Apart from the introduction, a thorough review of state-of-the-art multi-objective optimization (MO) techniques was presented in Chapter 2. This shows the way for the development of evolutionary theory, which enables the JG to further unravel the shortfall of MO, particularly when convergence and diversity are both in demand.

Considering the fact that the biogenetic phenomenon of the JG is new in computer science, this book needed to outline the basic biological gene transposition process to aid in comprehension of the concept. The transformation of gene manipulation in copy-and-paste and cut-and-paste processes into a computable language is crucial for gaining computationally executable coding. A full account of development in this area was provided in Chapter 3.

As the JG is so new in the domain of evolutionary algorithms, there was no previously reported literature. Thus, the task for deriving the necessary proofs of the JG mathematically was daunting. A number of theoretical proposals were considered, but only the schema theory deployed by Stephens and Waelbroeck's model [2] was finally adopted. This naturally led to the result in Chapter 4 of an exact mathematical formulation for the growth of the schemata rather than the concept of a lower-bound selection, as initially suggested by Holland [1].

As a result, two mathematical equations describing the growth of schemata were established, one for the copy-and-paste and the other for the cut-and-paste operation. These were reinforced by a thoughtful hypothesis for testing their evolutionary capability with a backup of purposefully designed statistical simulation runs. These results were then compared with the existing evolutionary operators, such as crossover, mutation, and random selection. It was then revealed that the JG was far more favorable for creating nondominated solutions and the subsequent solution for meeting the design goal quickly.

To reinforce this mathematical rhetoric so that the JG can stand firmly and be able to withstand the challenge from other evolutionary algorithms, relevant methods were devised for verifying these equations. The performance metrics for convergence and diversity were compared by introducing the method of the binary ε-indicator. All these were statistically conducted based on a range of benchmark mathematical functions, regardless of the forms of the constrained or unconstrained functions. Chapter 5 gave a full account of this exercise.

Based on all these successful work sequences, the natural progress would then be the ultimate engineering application in practice. The book pinpointed the relevant applications demonstrating the effectiveness of the JG, with particular emphasis on the conflicting issue of convergence versus diversity in the MO formulation.

Finally, it is good to know that the JG can now be a useful addition to the existing evolutionary algorithms. It also has a specific annotation: its ability to provide a quick converged solution as well as those solutions to outliers if necessary.

References

1. Holland, J. H., *Adaptation in Natural and Artificial Systems*, Ann Arbor, MI: MIT Press, 1975.
2. Stephens, C. R., Waelbroeck, H., Schemata evolution and building blocks, *Evolutionary Computation*, 7(2), 109–124, 1999.

Appendix A: Proofs of Lemmas in Chapter 4

Proof of Lemma 4.1

From Definition 4.4 in Chapter 4, the following identities can be easily obtained:

$$d(x_i, y_i) = d(x_j, x_i \oplus y_i \oplus x_j) \tag{A.1a}$$

$$d(x_i, *) = d(x_j, *) \tag{A.1b}$$

$$d(*, y_i) = d(*, x_i \oplus y_i \oplus x_j) \tag{A.1c}$$

where $x_i, y_i, x_j \in \{0,1\}$, and $\oplus$ is the logical XOR operator.

Based on Equation (A.1), it can be proved that

$$\Delta(\xi_i, V_k; \xi_m, G_{c,k}) = \Delta(\xi_i, V_k; \tilde{\xi}_m, G_{c,k}) \tag{A.2}$$

and

$$\Delta(\xi_i, V_k'; \xi_n, V_k') = \Delta(\xi_j, V_k'; \tilde{\xi}_n, V_k') \tag{A.3}$$

where

$$\tilde{\xi}_m(l) = \begin{cases} \xi_i(l') \oplus \xi_m(l) \oplus \xi_j(l') & l \in G_{c,k} \text{ and } \xi_m(l) \neq * \text{ and } \xi_i(l') \neq * \\ \xi_m(l) & \text{otherwise} \end{cases} \tag{A.4a}$$

and

$$\tilde{\xi}_n(l) = \begin{cases} \xi_i(l) \oplus \xi_n(l) \oplus \xi_j(l) & l \in V_k' \text{ and } \xi_n \neq * \\ \xi_n(l) & \text{otherwise} \end{cases} \tag{A.4b}$$

with $l' = k + (l - c)$. A directly visualized way to understand Equations (A.4) is shown in Figure A.1.

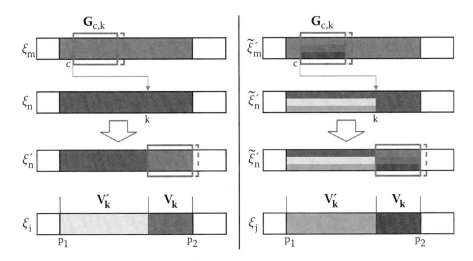

FIGURE A.1
Visualization of finding $\tilde{\xi}_m$ and $\tilde{\xi}_n$ (logical XOR is performed in striped areas).

Therefore, for $q = (\xi_m, \xi_n, c, k)$, there exists a unique $\tilde{q} = (\tilde{\xi}_m, \tilde{\xi}_n, c, k)$ such that

$$f^{(i)}(q) = \delta(f_T(\xi_i, V_k), f_T(\xi_m, G_{c,k})) \times \delta(f_T(\xi_i, V_k'), f_T(\xi_n, V_k'))$$

$$= \delta(f_T(\xi_j, V_k), f_T(\tilde{\xi}_m, G_{c,k})) \times \delta(f_T(\xi_j, V_k'), f_T(\tilde{\xi}_n, V_k'))$$

$$= f^j(\tilde{q})$$

Based on the logic specified in Equations (A.4), it can also be proved that if $f^{(i)}(q_1) = f^{(i)}(q_2) = f^{(j)}(\tilde{q})$, then $q_1 = q_2$. This completes the proof.

Proof of Lemma 4.2

Obviously, $\sum_{\xi_i \in S_\xi} P(\xi_i, t) = \sum_{\xi_i \in S_\xi} \frac{1}{M} = 1$ satisfies the constraint of $P(\xi_i, t)$. Assuming that $P(\xi_i, t) = \frac{1}{M}$, $\forall \xi_i \in S$, one has

$$P(\xi_i, t+1) = K \sum_{m=1}^{M} \sum_{n=1}^{M} \sum_{c=0}^{L-L_g} \sum_{k=0}^{L-L_g} \left[\delta(f_T(\xi_i, V_k), f_T(\xi_m, G_{c,k})) \times \delta(f_T(\xi_i, V_k'), f_T(\xi_n, V_k')) \right] \times \frac{1}{M^2}$$

$$= \frac{K}{M^2} \sum_{m=1}^{M} \sum_{n=1}^{M} \sum_{c=0}^{L-L_g} \sum_{k=0}^{L-L_g} \left[\delta(f_T(\xi_i, V_k), f_T(\xi_m, G_{c,k})) \times \delta(f_T(\xi_i, V_k'), f_T(\xi_n, V_k')) \right]$$

$$= \frac{K}{M^2} \sum_{q \in Q} f^{(i)}(q) \tag{A.5}$$

Using Lemma 4.1,

$$\sum_{q \in Q} f^{(i)}(q) = \sum_{\tilde{q} \in Q} f^{(j)}(\tilde{q}) \equiv C$$

(A.6)

Substituting Equation (A.6) into Equation (A.5), one has

$$P(\xi_i, t+1) = \frac{KC}{M^2}, \quad \forall \xi_i \in S$$

(A.7)

Since $\displaystyle\sum_{\xi_i \in S_\xi} P(\xi_i, t+1) = \sum_{\xi_i \in S} \frac{KC}{M^2} = \frac{KC}{M^2} \times M = 1$, 1 one has

$$C = \frac{M}{K}.$$

(A.8)

Substituting Equation (A.8) into Equation (A.6), one has

$$\sum_{q \in Q} f^{(i)}(q) = \frac{M}{K}.$$

(A.9)

Based on the definition, $\displaystyle\sum_{m=1}^{M}\sum_{n=1}^{M} a_{mn}^{(i)} = K \times \sum_{q \in Q} f^{(i)}(q) = K \times \frac{M}{K} = M$. This completes the proof.

Proof of Lemma 4.3

Consider

$$\sum_{i=1}^{M} a_{mn}^{(i)} = K \sum_{i=1}^{M} \sum_{c=0}^{L-L_g} \sum_{k=0}^{L-L_g} [\delta(f_T(\xi_i, V_k), f_T(\xi_m, G_{c,k})) \times \delta(f_T(\xi_i, V_k'), f_T(\xi_n, V_k'))]$$

$$= K \sum_{c=0}^{L-L_g} \sum_{k=0}^{L-L_g} \left\{ \sum_{i=1}^{M} [\delta(f_T(\xi_i, V_k), f_T(\xi_m, G_{c,k})) \times \delta(f_T(\xi_i, V_k'), f_T(\xi_n, V_k'))] \right\}$$

$$= K \sum_{c=0}^{L-L_g} \sum_{k=0}^{L-L_g} \left\{ \sum_{i=1}^{M} \left[\delta(f_T(\tilde{\xi}_i, V_k), f_T(\xi_p, G_{c,k})) \times \delta(f_T(\tilde{\xi}_i, V_k'), f_T(\xi_q, V_k')) \right] \right\}$$

(A.10)

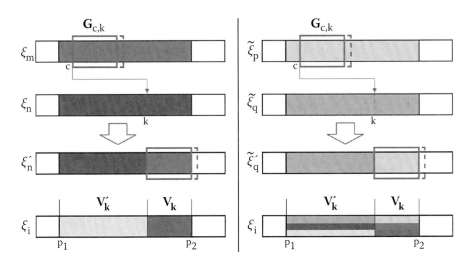

FIGURE A.2
Visualization of finding $\tilde{\xi}_i$ (logical XOR is performed in striped areas).

where

$$
\tilde{\xi}_i(l) = \begin{cases} \xi_i(l) \oplus \xi_m(l'') \oplus \xi_p(l''), & l \in V_k \text{ and } \xi_i(l) \neq * \text{ and } \xi_m(l') \neq * \\ \xi_i(l) \oplus \xi_n(l) \oplus \xi_q(l), & l \in V_k' \text{ and } \xi_i(l) \neq * \\ \xi_i(l), & \text{otherwise} \end{cases} \tag{A.11}
$$

where $l'' = c + (l - k)$. A way to understand Equation (A.11) visually is shown in Figure A.2.

Since m, n, p, and q are fixed, we have $\sum\limits_{i=1}^{M} a_{mn}^{(i)} = \sum\limits_{j=1}^{M} a_{pq}^{(j)}$, for any m, n, p, and q.

Therefore, $\sum\limits_{i=1}^{M} a_{mn}^{(i)} = C_1$ is a constant. From Lemma **4.2**, we know that

$$
\sum_{i=1}^{M} \left(\sum_{m=1}^{M} \sum_{n=1}^{M} a_{mn}^{(i)} \right) = \sum_{i=1}^{M} M = M^2. \tag{A.12}
$$

Meanwhile,

$$
\sum_{i=1}^{M} \left(\sum_{m=1}^{M} \sum_{n=1}^{M} a_{mn}^{(i)} \right) = \sum_{m=1}^{M} \sum_{n=1}^{M} \left(\sum_{i=1}^{M} a_{mn}^{(i)} \right) = \sum_{m=1}^{M} \sum_{n=1}^{M} C_1 = C_1 M^2 \tag{A.13}
$$

Therefore, $C_1 = 1$. This is equivalent to $\sum\limits_{i=1}^{M} A^{(i)} = \mathbf{1}_{M \times M}$.

Example A.1

Let $S_\xi = \{*00***,\ *01***,\ *10***,\ *11***\}$ and $L_g = 2$; we can obtain the following distribution matrices:

$$A^{(1)} = \begin{bmatrix} 0.78 & 0.24 & 0.24 & 0.1 \\ 0.64 & 0.14 & 0.14 & 0.04 \\ 0.64 & 0.14 & 0.14 & 0.04 \\ 0.54 & 0.08 & 0.08 & 0.02 \end{bmatrix}, \quad A^{(2)} = \begin{bmatrix} 0.1 & 0.64 & 0.04 & 0.18 \\ 0.16 & 0.66 & 0.06 & 0.16 \\ 0.16 & 0.66 & 0.06 & 0.16 \\ 0.18 & 0.64 & 0.04 & 0.1 \end{bmatrix},$$

$$A^{(3)} = \begin{bmatrix} 0.1 & 0.04 & 0.64 & 0.18 \\ 0.16 & 0.06 & 0.66 & 0.16 \\ 0.16 & 0.06 & 0.66 & 0.16 \\ 0.18 & 0.04 & 0.64 & 0.1 \end{bmatrix}, \quad A^{(4)} = \begin{bmatrix} 0.02 & 0.08 & 0.08 & 0.54 \\ 0.04 & 0.14 & 0.14 & 0.64 \\ 0.04 & 0.14 & 0.14 & 0.64 \\ 0.1 & 0.24 & 0.24 & 0.78 \end{bmatrix}.$$

It can be easily verified that

$$\sum_{m=1}^{4}\sum_{n=1}^{4} a_{mn}^{(1)} = 4,\ \sum_{m=1}^{4}\sum_{n=1}^{4} a_{mn}^{(2)} = 4,\ \sum_{m=1}^{4}\sum_{n=1}^{4} a_{mn}^{(3)} = 4,\ \text{and } \sum_{m=1}^{4}\sum_{n=1}^{4} a_{mn}^{(4)} = 4 \ \text{(Using Lemma 4.2)}$$

and

$$\sum_{i=1}^{4} A^{(i)} = \begin{bmatrix} 1 & 1 & 1 & 1 \\ 1 & 1 & 1 & 1 \\ 1 & 1 & 1 & 1 \\ 1 & 1 & 1 & 1 \end{bmatrix}. \ \text{(Using Lemma 4.3)}$$

Remark A.1

Let us consider

$$\Delta(\xi_i, V_k; \xi_m, G_{c,k}) = \prod_{j=0}^{m-1} d(v_1(j), v_2(j))$$

where $v_1 = f_T(\xi_i, V_k)$ and $v_2 = f_T(\xi_m, G_c)$. Based on Definition 4.4, we assume that there are $u_{c,k} + v_{c,k}$ actual bits in $v_1(j)$, where $u_{c,k}$ bits are compared with the actual bits in $v_2(j)$, and $v_{c,k}$ bits are compared with * in $v_2(j)$, as shown in Figure A.3. If all $u_{c,k}$ actual bits in $v_1(j)$ are the same as those in $v_2(j)$, we have

$$\Delta(\xi_i, V_k; \xi_m, G_{c,k}) = 2^{-v_{c,k}},$$

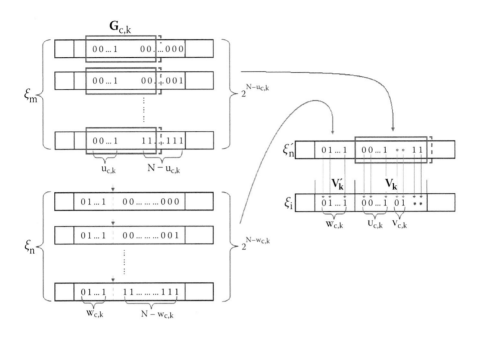

FIGURE A.3
Bit-to-bit comparison between schemata.

and $2^{N-u_{c,k}}$ of ξ_m in S_ξ fulfill this condition, where N is the order of ξ. As for the other $\xi_{m'}$ we have

$$\Delta(\xi_i, V_k; \xi_m, G_{c,k}) = 0.$$

Similarly, let us consider

$$\Delta(\xi_i, V'_k; \xi_n, V'_k) = \prod_{j=0}^{m-1} d(v_1(j), v_2(j)),$$

where $v_1 = f_T(\xi_i, V'_k)$, and $v_2 = f_T(\xi_n, V'_k)$. We assume that there are w_k actual bits in $v_1(j)$, as shown in Figure A.3. Apparently, $u_{c,k} + v_{c,k} + w_k = N$. It can be derived that 2^{N-w_k} of ξ_n in S_ξ gives

$$\Delta(\xi_i, V'_k; \xi_n, V'_k) = 1,$$

and the rest of ξ_n gives

$$\Delta(\xi_i, V'_k; \xi_n, V'_k) = 0.$$

Hence we can get

$$\sum_{m=1}^{M}\sum_{n=1}^{M}a_{mn}^{(i)} = K\sum_{m=1}^{M}\sum_{n=1}^{M}\sum_{c=0}^{L-L_g}\sum_{k=0}^{L-L_g}\Delta(\xi_i,V_k;\xi_m,G_{c,k})\times\Delta(\xi_i,V_k';\xi_n,V_k')$$

$$= K\sum_{c=0}^{L-L_g}\sum_{k=0}^{L-L_g}\left[\left(\sum_{m=1}^{M}\Delta(\xi_i,V_k;\xi_m,G_{c,k})\right)\times\left(\sum_{n=1}^{M}\Delta(\xi_i,V_k';\xi_n,V_k')\right)\right]$$

$$= K\sum_{c=0}^{L-L_g}\sum_{k=0}^{L-L_g}\left[\left(2^{N-u_{c,k}}\times 2^{-v_{c,k}}\right)\times\left(2^{N-w_k}\times 1\right)\right]$$

$$= K\sum_{c=0}^{L-L_g}\sum_{k=0}^{L-L_g}\left[2^{2N-u_{c,k}-v_{c,k}-w_k}\right]$$

$$= K\sum_{c=0}^{L-L_g}\sum_{k=0}^{L-L_g}\left[2^{N}\right]$$

$$= M$$

This is actually another way to prove Lemma 4.2.
  Apparently,

$$a_{mi}^{(i)} = K\sum_{c=0}^{L-L_g}\sum_{k=0}^{L-L_g}\Delta(\xi_i,V_k;\xi_m,G_{c,k})\times\Delta(\xi_i,V_k';\xi_i,V_k')$$

$$= K\sum_{c=0}^{L-L_g}\sum_{k=0}^{L-L_g}\Delta(\xi_i,V_k;\xi_m,G_{c,k})\times 1 \qquad\qquad\text{(A.14)}$$

$$\geq K\sum_{c=0}^{L-L_g}\sum_{k=0}^{L-L_g}\Delta(\xi_i,V_k;\xi_m,G_{c,k})\times\Delta(\xi_i,V_k';\xi_n,V_k')$$

$$= a_{mn}^{(i)}$$

Hence, for each row of the distribution matrix $A^{(i)}$, the element in the *i*th column has the largest value.

If we sum all elements in the *i*th column of the distribution matrix $A^{(i)}$, we can get

$$\sum_{m=1}^{M} a_{mi}^{(i)} = K \sum_{m=1}^{M} \sum_{c=0}^{L-L_g} \sum_{k=0}^{L-L_g} \Delta(\xi_i, V_k; \xi_m, G_{c,k}) \times \Delta(\xi_i, V_k'; \xi_i, V_k')$$

$$= K \sum_{c=0}^{L-L_g} \sum_{k=0}^{L-L_g} \left(\sum_{m=1}^{M} \Delta(\xi_i, V_k; \xi_m, G_{c,k}) \times 1 \right)$$

$$= K \sum_{c=0}^{L-L_g} \sum_{k=0}^{L-L_g} \left(2^{N-u_{c,k}} \times 2^{-v_{c,k}} \right)$$

(A.15)

$$= K \sum_{c=0}^{L-L_g} \sum_{k=0}^{L-L_g} \left(2^{N-u_{c,k}-v_{c,k}} \right)$$

$$= K \sum_{c=0}^{L-L_g} \sum_{k=0}^{L-L_g} \left(2^{w_k} \right)$$

$$= \frac{1}{L-L_g+1} \sum_{k=0}^{L-L_g} 2^{w_k}$$

It is easy to prove that $\sum_{m=1}^{M} a_{mi}^{(i)}$ decreases when L_g increases.

Proof of Lemma 4.4

Based on the definition of $a_{mn}^{(i)}$ given in Equation (4.54), consider

$$\sum_{\xi_n \in Z} a_{mn}^{(i)} - \sum_{\xi_n \in Z'} a_{mn}^{(i)} = K \sum_{c=0}^{L-L_g} \sum_{k=0}^{L-L_g} \Delta(\xi_i, V_k; \xi_m, G_{c,k})$$

(A.16)

$$\times \left[\sum_{\xi_n \in Z} \Delta(\xi_i, V_k'; \xi_n, V_k') - \sum_{\xi_n \in Z'} \Delta(\xi_i, V_k'; \xi_n, V_k') \right]$$

and for any particular $\xi_n \in Z$ (note: Z can be Z_{odd} or Z_{even}), we have the following two different cases:

a. For any pasting position k, which makes some actual bits of ξ_i be located in V_k, it is possible to flip the rightmost bit in V_k to find a corresponding $\xi_{n'} \in Z'$ such that

$$\Delta(\xi_i, V_k'; \xi_n, V_k') = \Delta(\xi_i, V_k'; \xi_{n'}, V_k').$$

Hence, if there are some actual bits of ξ_i located in V_k, the difference term in Equation (A.16) is zero, that is,

$$\left[\sum_{\xi_n \in Z} \Delta(\xi_i, V_k'; \xi_n, V_k') - \sum_{\xi_n \in Z'} \Delta(\xi_i, V_k'; \xi_n, V_k') \right] = 0$$

b. For any pasting position k, which makes all the actual bits be located in V_k' only when $\xi_n = \xi_i$, $\Delta(\xi_i, V_k'; \xi_n, V_k') = 1$; otherwise, $\Delta(\xi_i, V_k'; \xi_n, V_k') = 0$. We define a set K':

$$K' = \{k: \text{no actual bit in pasting region}\}.$$

For any $k \in K'$, we have

$$\Delta(\xi_i, V_k; \xi_m, G_{c,k}) = 1$$

and

$$\left[\sum_{\xi_n \in Z} \Delta(\xi_i, V_k'; \xi_n, V_k') - \sum_{\xi_n \in Z'} \Delta(\xi_i, V_k'; \xi_n, V_k') \right] = \begin{cases} 1, & \xi_i \in Z \\ -1, & \xi_i \in Z' \end{cases}$$

Figure A.4 helps clarify cases a and b.
By summing cases a and b, we have

$$\sum_{\xi_n \in Z} a_{mn}^{(i)} - \sum_{\xi_n \in Z'} a_{mn}^{(i)} = \pm K \sum_{c=0}^{L-L_g} \sum_{k \in K'} 1$$

$$= \pm \frac{N_0}{L - L_g - 1}$$

$$\equiv C_{row}$$

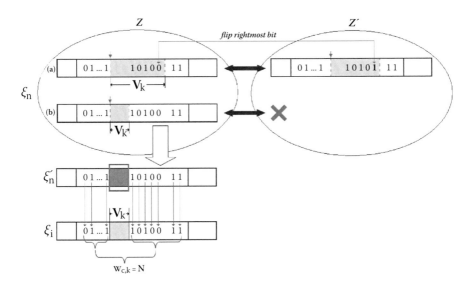

FIGURE A.4
Visualization of the proof of Lemma 4.4.

where N_0 is the size of K', specifying the number of possible pasting positions in ξ_i that make the pasted region contain don't-care bits (*) only. If $\xi_i \in Z$, C_{row} is positive or zero; otherwise, C_{row} is negative or zero. Thus, the proof is completed.

Proof of Lemma 4.5

Based on the definition of $a_{mn}^{(i)}$ given in Equation (4.54), consider

$$\sum_{\xi_m \in Z} a_{mn}^{(i)} - \sum_{\xi_m \in Z'} a_{mn}^{(i)} = K \sum_{c=0}^{L-L_g} \sum_{k=0}^{L-L_g} \Delta(\xi_i, V_k'; \xi_n, V_k')$$

$$\times \left[\sum_{\xi_m \in Z} \Delta(\xi_i, V_k; \xi_m, G_{c,k}) - \sum_{\xi_m \in Z'} \Delta(\xi_i, V_k; \xi_m, G_{c,k}) \right] \qquad (A.17)$$

and for any particular $\xi_m \in Z$ (note: Z can be Z_{odd} or Z_{even}), as shown in Figure A.5, three cases can be obtained as follows:

a. For any copying position c, which causes some actual bits of ξ_m to be located outside $G_{c,k}$, it is possible to flip the rightmost of those bits to find a corresponding $\xi_{m'} \in Z'$ such that

$$\Delta(\xi_i, V_k; \xi_m, G_{c,k}) = \Delta(\xi_i, V_k; \xi_{m'}, G_{c,k}).$$

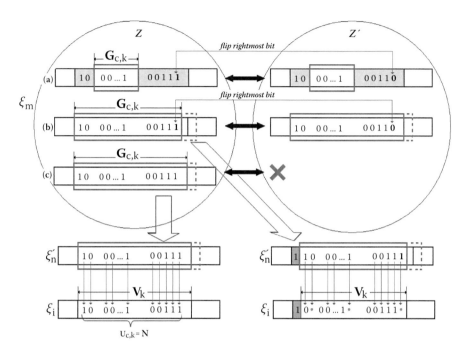

FIGURE A.5
Visualization of the proof of Lemma 4.5.

Hence, if some actual bits are not copied in the transposon, we have

$$\left[\sum_{\xi_m \in Z} \Delta(\xi_i, V_k; \xi_m, G_{c,k}) - \sum_{\xi_m \in Z'} \Delta(\xi_i, V_k; \xi_m, G_{c,k}) \right] = 0.$$

b. For any copying position c, which causes all the actual bits of ξ_m to be located inside $G_{c,k}$, if $k \neq c$, that is, any actual bit in v_2 is compared with the don't-care bit in v_1, it is possible to find a corresponding $\xi_{m'} \in Z'$ by flipping the rightmost of those actual bits, such that

$$\Delta(\xi_i, V_k; \xi_m, G_{c,k}) = \Delta(\xi_i, V_k; \xi_{m'}, G_{c,k}).$$

Hence, if all actual bits are copied in the transposon and some of those are compared with don't-care bits after they are pasted,

$$\left[\sum_{\xi_m \in Z} \Delta(\xi_i, V_k; \xi_m, G_{c,k}) - \sum_{\xi_m \in Z'} \Delta(\xi_i, V_k; \xi_m, G_{c,k}) \right] = 0.$$

c. For any copying position c, which causes all the actual bits of ξ_m to be located inside $G_{c,k}$, if $k = c$, all actual bits in v_2 are compared with actual bits in v_1. Only when $\xi_m = \xi_i$, $\Delta(\xi_i, V_k; \xi_m, G_{c,k}) = 1$; otherwise, $\Delta(\xi_i, V_k; \xi_m, G_{c,k}) = 0$. We have $\Delta(\xi_i, V'_k; \xi_n, V'_k) = 1$ and

$$
\left[\sum_{\xi_m \in Z} \Delta(\xi_i, V_k; \xi_m, G_{c,k}) - \sum_{\xi_m \in Z'} \Delta(\xi_i, V_k; \xi_m, G_{c,k}) \right] = \begin{cases} 1, & \xi_i \in Z \\ -1, & \xi_i \in Z' \end{cases}
$$

By summing cases a to c,

$$
\sum_{\xi_m \in Z} a_{mn}^{(i)} - \sum_{\xi_m \in Z'} a_{mn}^{(i)} = \pm K \sum_{c \in C'} \sum_{k=c} 1
$$

$$
= \pm \frac{1}{(L - L_g - 1)^2} \sum_{c \in C'} 1
$$

$$
= \pm \frac{N_{all}}{(L - L_g - 1)^2}
$$

$$
\equiv C_{col}
$$

where the set C' is defined as $C' = \{c : \text{all actual bits are inside } G_{c,k}\}$, and N_{all} is the size of C', that is, the number of possible copying positions such that the transposon region in ξ_i includes all the actual bits or the number of possible pasting positions such that the pasted region in ξ_i includes all the actual bits. If $\xi_i \in Z$, C_{col} is positive or zero; otherwise, C_{col} is negative or zero.

Example A.2

Let $S_\xi = \{*00***, \ *01***, \ *10***, \ *11***\}$ and $L_g = 2$; we can obtain the following distribution matrix for ξ_1:

$$
A = \begin{bmatrix} 0.78 & 0.24 & 0.24 & 0.1 \\ 0.64 & 0.14 & 0.14 & 0.04 \\ 0.64 & 0.14 & 0.14 & 0.04 \\ 0.54 & 0.08 & 0.08 & 0.02 \end{bmatrix}.
$$

By adding columns 1 and 4 and then subtracting columns 2 and 3, we can get

$$
\left(
\begin{bmatrix} 0.78 \\ 0.64 \\ 0.64 \\ 0.54 \end{bmatrix}
+
\begin{bmatrix} 0.1 \\ 0.04 \\ 0.04 \\ 0.02 \end{bmatrix}
\right)
-
\left(
\begin{bmatrix} 0.24 \\ 0.14 \\ 0.14 \\ 0.08 \end{bmatrix}
+
\begin{bmatrix} 0.24 \\ 0.14 \\ 0.14 \\ 0.08 \end{bmatrix}
\right)
=
\begin{bmatrix} 0.4 \\ 0.4 \\ 0.4 \\ 0.4 \end{bmatrix}.
$$

as there are two positions that contain don't-care bits only (i.e., *00*** and *00***). Thus

$$
N_0 = 2, \text{ and } C_{row} = \frac{N_0}{L - L_g + 1} = \frac{2}{5} = 0.4. \text{ (Using Lemma 4.4)}
$$

By adding rows 1 and 4 and then subtracting rows 2 and 3, we can get

$$
([0.78 \quad 0.24 \quad 0.24 \quad 0.1] + [0.54 \quad 0.08 \quad 0.08 \quad 0.02])
$$
$$
- ([0.64 \quad 0.14 \quad 0.14 \quad 0.04] + [0.64 \quad 0.14 \quad 0.14 \quad 0.04])
$$
$$
= [0.04 \quad 0.04 \quad 0.04 \quad 0.04]
$$

as there is only a single position that contains all the actual bits (i.e., *00***). Thus

$$
N_{all} = 1, \text{ and } C_{col} = \frac{N_{all}}{\left(L - L_g + 1\right)^2} = \frac{1}{25} = 0.04. \text{ (Using Lemma 4.5)}
$$

Proof of Lemma 4.6

If $\xi_i \in Z$, C_{row} and C_{col} are both nonnegative. Otherwise, they are both nonpositive.

Apparently,

$$
0 \le N_0 + N_{all} \le L - L_g + 1
$$
$$
0 \le \frac{N_0}{L - L_g + 1} + \frac{N_{all}}{L - L_g + 1} \le 1
$$

Since

$$0 \le \frac{N_{all}}{(L - L_g + 1)^2} < \frac{N_{all}}{L - L_g + 1} \quad \text{for } L_g < L,$$

one has

$$0 \le \frac{N_0}{L - L_g + 1} + \frac{N_{all}}{(L - L_g + 1)^2} < 1, \text{ that is, } -1 < C_{row} + C_{col} < 1.$$

Proof of Lemma 4.7

Based on the row and column properties given in Lemmas **4.4** and **4.5**, one has

$$Q_1 - Q_3 = \frac{M}{2} C_{row},$$

$$Q_1 - Q_4 = \frac{M}{2} C_{col},$$

$$Q_3 - Q_2 = \frac{M}{2} C_{col},$$

and with Lemma **4.2**, we can get

$$\sum_{i=1}^{4} Q_i = M.$$

The proof can be completed by solving the equations directly.
For illustration, the elements of Q_1, Q_2, Q_3, Q_4 for $M = 2$ and $M = 4$ are depicted in Figure A.6.

Proof of Lemma 4.8

The proof for Lemma 4.8 is derived in the same way as that of Lemma **4.2**.

Proof of Lemma 4.9

The proof of Lemma 4.9 is derived in the same way as that of Lemma **4.3**.

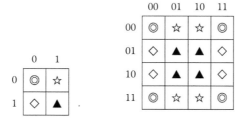

FIGURE A.6
Classification of elements in a distribution matrix ($\odot,\star,\diamond,\blacktriangle$ represent $a_{mn}^{(i)}$ with $\xi_m,\xi_n \in Z$; $\xi_m \in Z, \xi_n \in Z', \xi_m \in Z', \xi_n \in Z$; and $\xi_m, \xi_n \in Z'$, respectively).

Proof of Lemma 4.10

Based on the definition of $b_{mn}^{(i)}$ given in Equation (4.60), consider

$$\sum_{\xi_n \in Z} b_{mn}^{(i)} - \sum_{\xi_n \in Z'} b_{mn}^{(i)} = \frac{1}{(L-L_g+1)^3} \sum_{c_m=0}^{L-L_g} \sum_{c_n=0}^{L-L_g} \sum_{k_n \notin \kappa(c_n)} \Delta(\xi_i, I_{c_n,k_n}; \xi_m, G_{c_m,c_n,k_n})$$

$$\times \left[\sum_{\xi_n \in Z} \Delta(\xi_i, M_{c_n,k_n}; \xi_n, M'_{c_n,k_n})\Delta(\xi_i, R_{c_n,k_n}; \xi_n, R_{c_n,k_n}) \right.$$

$$\left. - \sum_{\xi_n \in Z'} \Delta(\xi_i, M_{c_n,k_n}; \xi_n, M'_{c_n,k_n})\Delta(\xi_i, R_{c_n,k_n}; \xi_n, R_{c_n,k_n}) \right]$$

(A.18)

where $\kappa(c_n) = (c_n, c_n + L_g]$, and for any particular $\xi_n \in Z$ (note: Z can be Z_{odd} or Z_{even}), the following three cases can be obtained:

a. For any cutting position c_n, which cuts some actual bits in ξ_n, it is possible to flip the rightmost of those bits to find a corresponding $\xi_{n'} \in Z'$, such that the difference term in Equation (A.18) is zero, that is,

$$\sum_{\xi_n \in Z} \Delta(\xi_i, M_{c_n,k_n}; \xi_n, M'_{c_n,k_n})\Delta(\xi_i, R_{c_n,k_n}; \xi_n, R_{c_n,k_n})$$

$$- \sum_{\xi_n \in Z'} \Delta(\xi_i, M_{c_n,k_n}; \xi_n, M'_{c_n,k_n})\Delta(\xi_i, R_{c_n,k_n}; \xi_n, R_{c_n,k_n}) = 0$$

b. For any cutting position c_n, which cuts no actual bits in ξ_n, and for any pasting position k_n, if some actual bits in M'_{c_n,k_n} are compared with don't-care bits in M_{c_n,k_n} of ξ_i, it is possible to flip the rightmost of those actual bits to find a corresponding $\xi_{n'} \in Z'$ such that

$$\sum_{\xi_n \in Z} \Delta(\xi_i, M_{c_n,k_n}; \xi_n, M'_{c_n,k_n}) \Delta(\xi_i, R_{c_n,k_n}; \xi_n, R_{c_n,k_n})$$

$$- \sum_{\xi_n \in Z'} \Delta(\xi_i, M_{c_n,k_n}; \xi_n, M'_{c_n,k_n}) \Delta(\xi_i, R_{c_n,k_n}; \xi_n, R_{c_n,k_n}) = 0$$

c. For any cutting position c_n, which cuts no actual bits in ξ_n, and for any pasting position k_n, if all actual bits in M'_{c_n,k_n} are compared with actual bits in M_{c_n,k_n} of ξ_i, it is easy to verify that there are no actual bits in M'_{c_n,k_n}. So, all actual bits are located inside R_{c_n,k_n}. We define a set

$$S_{ck} = \{(c_n, k_n): \text{ no actual bits in cutting and moving regions}\}.$$

For any pair of $(c_n, k_n) \in S_{ck}$, we have

$$\Delta(\xi_i, I_{c_n,k_n}; \xi_m, G_{c_m,c_n,k_n}) = 1$$

and

$$\sum_{\xi_n \in Z} \Delta\left(\xi_i, M_{c_n,k_n}; \xi_n, M'_{c_n,k_n}\right) \Delta(\xi_i, R_{c_n,k_n}; \xi_n, R_{c_n,k_n})$$

$$- \sum_{\xi_n \in Z'} \Delta\left(\xi_i, M_{c_n,k_n}; \xi_n, M'_{c_n,k_n}\right) \Delta(\xi_i, R_{c_n,k_n}; \xi_n, R_{c_n,k_n}) = \pm 1$$

By summing cases a to c, we get

$$\sum_{\xi_n \in Z} b_{mn}^{(i)} - \sum_{\xi_n \in Z'} b_{mn}^{(i)} = \pm \frac{1}{(L-L_g+1)^3} \sum_{c_m=0}^{L-L_g} \sum_{(c_n,k_n) \in S_{ck}} 1$$

$$= \pm \frac{N'_0}{(L-L_g+1)^2}$$

$$\equiv C'_{row}$$

where N_0' is the size of S_{ck}, that is, the number of possible combinations of the cutting and pasting positions such that the transposon and moving regions in ξ_i only contain the don't-care bit (*). If $\xi_i \in Z$, C_{row}' is positive or zero; otherwise, it is negative or zero.

Proof of Lemma 4.11

Based on the definition of $b_{mn}^{(i)}$ given in Equation (4.60), consider

$$
\sum_{\xi_m \in Z} b_{mn}^{(i)} - \sum_{\xi_m \in Z'} b_{mn}^{(i)}
$$

$$
= \frac{1}{(L - L_g + 1)^3} \sum_{c_m=0}^{L-L_g} \sum_{c_n=0}^{L-L_g} \sum_{k_n \notin \kappa(c_n)} \Delta\left(\xi_i, M_{c_n,k_n}; \xi_n, M_{c_n,k_n}'\right) \Delta\left(\xi_i, R_{c_n,k_n}; \xi_n, R_{c_n,k_n}\right) \quad \text{(A.19)}
$$

$$
\times \left[\sum_{\xi_m \in Z} \Delta\left(\xi_i, I_{c_n,k_n}; \xi_m, G_{c_m,c_n,k_n}\right) - \sum_{\xi_m \in Z'} \Delta\left(\xi_i, I_{c_n,k_n}; \xi_m, G_{c_m,c_n,k_n}\right) \right]
$$

where $\kappa(c_n) = (c_n, c_n + L_g]$.

Again, for any particular $\xi_m \in Z$ (Z_{odd} or Z_{even}), the following three cases are obtained:

a. For any cutting position c_m, which leaves some actual bits outside G_{c_m,c_n,k_n} in ξ_m, it is possible to flip the rightmost of those actual bits to find a corresponding $\xi_{m'} \in Z'$, such that

$$
\sum_{\xi_m \in Z} \Delta\left(\xi_i, I_{c_n,k_n}; \xi_m, G_{c_m,c_n,k_n}\right) - \sum_{\xi_m \in Z'} \Delta\left(\xi_i, I_{c_n,k_n}; \xi_m, G_{c_m,c_n,k_n}\right) = 0
$$

b. For any cutting position c_m, which cuts all the actual bits of ξ_m inside G_{c_m,c_n,k_n}, and for any pasting position k_n, if some actual bits in G_{c_m,c_n,k_n} are compared with don't-care bits in I_{c_n,k_n} of ξ_i, it is possible to flip the rightmost of those actual bits to find a corresponding $\xi_{m'} \in Z'$, such that

$$
\sum_{\xi_m \in Z} \Delta\left(\xi_i, I_{c_n,k_n}; \xi_m, G_{c_m,c_n,k_n}\right) - \sum_{\xi_m \in Z'} \Delta\left(\xi_i, I_{c_n,k_n}; \xi_m, G_{c_m,c_n,k_n}\right) = 0
$$

c. For any cutting position c_m, which cuts all the actual bits of ξ_m inside G_{c_m,c_n,k_n}, and for any pasting position k_n, if all actual bits in G_{c_m,c_n,k_n} are compared with actual bits in I_{c_n,k_n} of ξ_i, it is easy to verify that the region G_{c_m,c_n,k_n} overlaps the region I_{c_n,k_n}. We have

$$\Delta\left(\xi_i, M_{c_n,k_n}; \xi_n, M'_{c_n,k_n}\right)\Delta(\xi_i, R_{c_n,k_n}; \xi_n, R_{c_n,k_n}) = 1$$

and

$$\sum_{\xi_m \in Z} \Delta\left(\xi_i, I_{c_n,k_n}; \xi_m, G_{c_m,c_n,k_n}\right) - \sum_{\xi_m \in Z'} \Delta(\xi_i, I_{c_n,k_n}; \xi_m, G_{c_m,c_n,k_n}) = \pm 1$$

By summing all three cases, we get

$$\sum_{\xi_m \in Z} b_{mn}^{(i)} - \sum_{\xi_m \in Z'} b_{mn}^{(i)} = \pm \frac{1}{(L-L_g+1)^3} \sum_{c_n=0}^{L-L_g} \sum_{c_m \in C'} \sum_{k_n=c_n \text{ or } c_n+L_g} 1$$

$$= \pm \frac{1}{(L-L_g+1)^2} \sum_{c_m \in C'} 1$$

$$= \pm \frac{N_{all}}{(L-L_g+1)^2}$$

$$\equiv C'_{col}$$

where C' and N_{all} follow the same definition as in Lemma 4.5. If $\xi_i \in Z$, C'_{col} is positive or zero; otherwise, it is negative or zero.

Proof of Lemma 4.12

Apparently,

$$0 \le N'_0 + N_{all} < (L-L_g+1)^2$$

$$0 \le \frac{N'_0}{(L-L_g+1)^2} + \frac{N_{all}}{(L-L_g+1)^2} < 1$$

Therefore, $-1 < C'_{row} + C'_{col} < 1$.

Appendix B: Benchmark Test Functions

The unconstrained and constrained test functions used in our study are provided in Tables B.1 and B.2, respectively. The natures of true Pareto-optimal fronts of the test functions are also shown in the first columns of each table. Moreover, the true Pareto-optimal sets and the corresponding true Pareto-optimal fronts of both unconstrained and constrained test functions utilized are depicted in Figures B.1, B.2, and B.3.

TABLE B.1

Unconstrained Test Functions

Test Functions (Nature of True Pareto Front)	Total Number of Variables n	Variable Bounds	Objective Functions
SCH (convex)	1	$[-10^5, 10^5]$	$f_1(x) = x^2$ $f_2(x) = (x-2)^2$
FON (concave)	3	$[-4, 4]$	$f_1(x) = 1 - \exp\left(-\sum_{i=1}^{n}\left(x_i - \frac{1}{\sqrt{n}}\right)^2\right)$ $f_2(x) = 1 - \exp\left(-\sum_{i=1}^{n}\left(x_i + \frac{1}{\sqrt{n}}\right)^2\right)$
POL (convex and disconnected)	2	$[-\pi, \pi]$	$f_1(x) = 1 + (A_1 - B_1)^2 + (A_2 - B_2)^2$ $f_2(x) = (x_1 + 3)^2 + (x_2 + 1)^2$ $A_1 = 0.5\sin 1 - 2\cos 1 + \sin 2 - 1.5\cos 2$ $A_2 = 1.5\sin 1 - \cos 1 + 2\sin 2 - 0.5\cos 2$ $B_1 = 0.5\sin x_1 - 2\cos x_1 + \sin x_2 - 1.5\cos x_2$ $B_2 = 1.5\sin x_1 - \cos x_1 + 2\sin x_2 - 0.5\cos x_2$
ZIT1 (convex)	30	$[0, 1]$	$f_1(x) = x_1$ $f_2(x) = g(x)\left[1 - \sqrt{\frac{x_1}{g(x)}}\right]$ $g(x) = 1 + \frac{9 \cdot \sum_{i=2}^{n} x_i}{n-1}$

(continued)

TABLE B.1 (CONTINUED)

Unconstrained Test Functions

Test Functions (Nature of True Pareto Front)	Total Number of Variables n	Variable Bounds	Objective Functions
ZIT2 (concave)	30	[0, 1]	$f_1(x) = x_1$
			$f_2(x) = g(x)\left[1 - \left(\dfrac{x_1}{g(x)}\right)^2\right]$
			$g(x) = 1 + \dfrac{9 \cdot \Sigma_{i=2}^{n} x_i}{n-1}$
ZIT3 (convex and disconnected)	30	[0, 1]	$f_1(x) = x_1$
			$f_2(x) = g(x)\left[1 - \sqrt{\dfrac{x_1}{g(x)}} - \dfrac{x_1}{g(x)} \sin 10\pi x_1\right]$
			$g(x) = 1 + \dfrac{9 \cdot \Sigma_{i=2}^{n} x_i}{n-1}$
ZIT4 (convex)	10	$x_1 \in [0, 1,]$ and $x_i \in [-5, 5]$ for $i = 2, 3, \cdots, n$	$f_1(x) = x_1$ $f_2(x) = g(x)\left[1 - \sqrt{\dfrac{x_1}{g(x)}}\right]$ $g(x) = 1 + 10(n-1) + \Sigma_{i=2}^{n}\left(x_i^2 - 10\cos 4\pi x_i\right)$
ZIT6 (concave and nonuniformly spaced)	10	[0, 1]	$f_1(x) = 1 - \exp(-4x_1)\sin^6 6\pi x_1$
			$f_2(x) = g(x)\left[1 - \left(\dfrac{f_1(x)}{g(x)}\right)^2\right]$
			$g(x) = 1 + 9\left(\dfrac{\Sigma_{i=2}^{n} x_i}{n-1}\right)^{0.25}$

Source: Data from Chan, T. M., Man, K. F., Kwong, S., Tang, K. S., A jumping gene paradigm for evolutionary multiobjective optimization, *IEEE Transactions on Evolutionary Computation*, 12(2), 143–159, 2008.

TABLE B.2

Constrained Test Functions

Test Functions (Nature of True Pareto Front)	Variable Bounds	Objective Functions	Constraints
DEB (connected)	$x_1 \in [0.1, 1.0]$ $x_2 \in [0.5]$	$f_1(x) = x_1$ $f_2(x) = \dfrac{1 + x_2}{x_1}$	$g_1(x) = x_2 + 9x_1 \geq 6$ $g_2(x) = -x_2 + 9x_1 \geq 1$
BEL (connected)	$x_1 \in [0, 5]$ $x_2 \in [0, 3]$	$f_1(x) = -2x_1 + x_2$ $f_2(x) = 2x_1 + x_2$	$g_1(x) = -x_1 + x_2 \leq 1$ $g_2(x) = x_1 + x_2 \leq 7$
SRIN (connected)	$x_i \in [-20, 20]$ for $i = 1, 2$	$f_1(x) = (x_1 - 2)^2 + (x_2 - 1)^2 + 2$ $f_2(x) = 9x_1 - (x_2 - 1)^2$	$g_1(x) = x_1^2 + x_2^2 \leq 225$ $g_2(x) = x_1 - 3x_2 \leq -10$
TAN (disconnected and convoluted)	$x_i \in [0, \pi]$ for $i = 1, 2$	$f_1(x) = x_1$ $f_2(x) = x_2$	$g_1(x) = -x_1^2 - x_2^2 + 1$ $\quad + 0.1\cos\left[16\arctan\left(\dfrac{x_1}{x_2}\right)\right] \leq 0$ $g_2(x) = (x_1 - 0.5)^2$ $\quad + (x_2 - 0.5)^2 \leq 0.5$
BINH (convex)	$x_i \in [-15, 30]$ for $i = 1, 2$	$f_1(x) = 4x_1^2 + 4x_2^2$ $f_2(x) = (x_1 - 5)^2 + (x_2 - 5)^2$	$g_1(x) = (x_1 - 5)^2 + x_2^2 \leq 25$ $g_2(x) = -(x_1 - 8)^2 + (x_2 + 3)^2 \leq -7.7$

Source: Data from Chan, T. M., Man, K. F., Kwong, S., Tang, K. S., A jumping gene paradigm for evolutionary multiobjective optimization, *IEEE Transactions on Evolutionary Computation*, 12(2), 143–159, 2008.

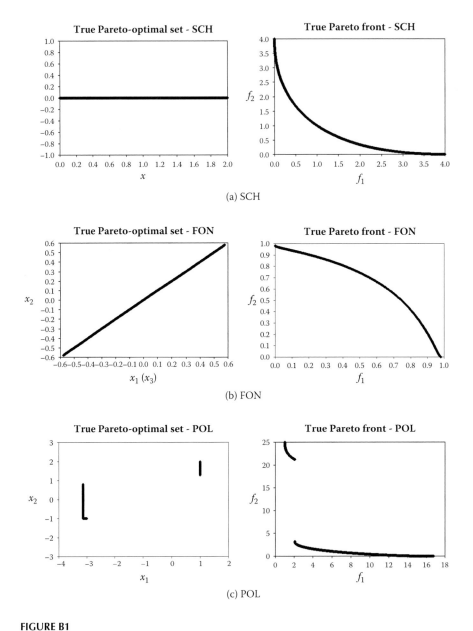

FIGURE B1
The true Pareto-optimal sets and corresponding true Pareto fronts of unconstrained test functions SCH, FON, and POL.

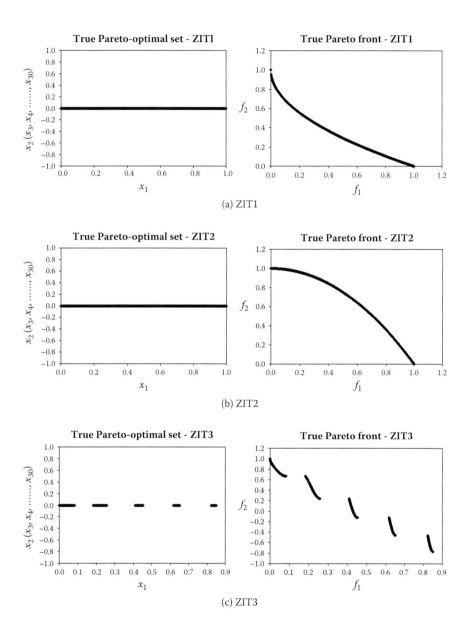

FIGURE B2
The true Pareto-optimal sets and corresponding true Pareto fronts of unconstrained test functions ZIT1, ZIT2, ZIT3, ZIT4, and ZIT6.

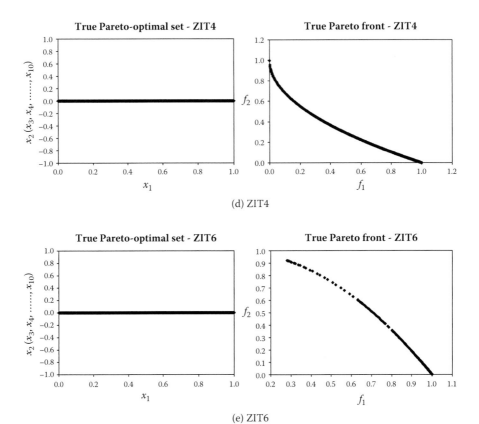

(d) ZIT4

(e) ZIT6

FIGURE B2 (CONTINUED)

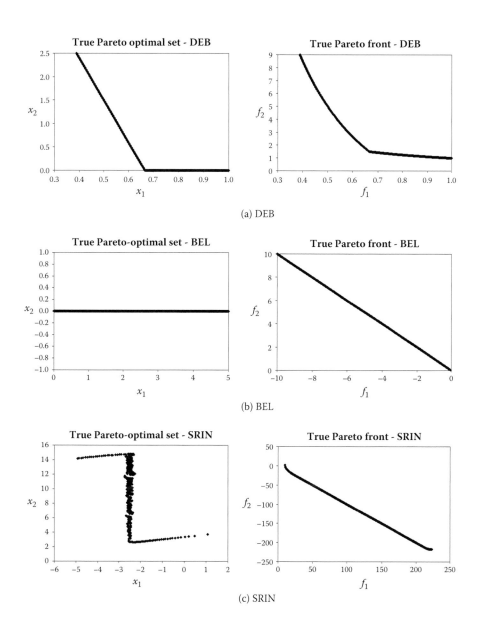

FIGURE B3
The true Pareto-optimal sets and corresponding true Pareto fronts of constrained test functions.

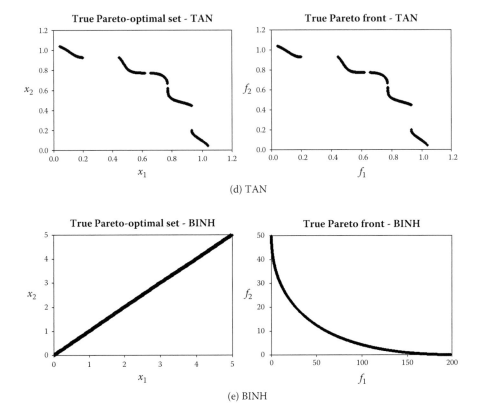

(d) TAN

(e) BINH

FIGURE B3 (CONTINUED)

Reference

1. Chan, T. M., Man, K. F., Kwong, S., Tang, K. S., A jumping gene paradigm for evolutionary multiobjective optimization, *IEEE Transactions on Evolutionary Computation*, 12(2), 143–159, 2008.

Appendix C: Chromosome Representation

Two types of chromosome representations, real and binary, were adopted for the evaluation of test functions. Regarding the real representation, the encoding and decoding of genes are straightforward. One decision variable can be directly encoded to a gene, and this gene can be decoded back to the original decision variable with no extra effort. But a method of determining the total length of a chromosome and a decoding function are required for binary representation [2]. Their details are given next.

First, let us consider the former method and assume that there are two decision variables in an optimization problem, and both of their bounds are $[-4, 4]$. If the required precision is six decimal places, the range $[-4, 4]$ should be divided into at least $8 \times 1000000 + 1$ equal size ranges. Since $2^{22} < 8,000,001 < 2^{23}$, it turns out that 23 bits are needed to represent one decision variable, and the total length of a chromosome for encoding two decision variables is 46 bits.

On the other hand, the decoding function starts with converting a binary string $(b_{n-1} b_{n-2} \ldots b_0)$ to the corresponding decimal value a using the formula

$$a = \sum_{i=0}^{n-1} b_i \cdot 2^i, \quad \forall b_i \in \{0, 1\} \tag{C.1}$$

where n is the total number of bits in the string.

Then, the value of a decision variable x can be obtained by

$$x = B_L + \frac{a}{2^n - 1} \cdot (B_U - B_L) \tag{C.2}$$

where B_L and B_U are the lower and upper bounds of the decision variable, respectively.

Consider an example of a binary string (10100001000100001001001) with $n = 23$ bits; the value of a is 5277769. Assuming that $B_L = -4$ and $B_U = 4$, the value of the decision variable is

$$x = -4 + \frac{5277769}{2^{23} - 1} \cdot (4 - (-4)) \approx 1.0333$$

TABLE C.1

Total Length of a Chromosome for Each Unconstrained and Constrained Test Function in the Binary Chromosome Representation

Test Function	Total Number of Variables	Total Length of a Chromosome (bits)
SCH	1	38
FON	3	69
POL	2	46
ZIT1, ZIT2, ZIT3	30	600
ZIT4	10	236
ZIT6	10	200
DEB	2	43
BEL	2	45
SRIN, BINH	2	52
TAN	2	44

Source Data from Chan, T. M., Man, K. F., Kwong, S., Tang, K. S., A jumping gene paradigm for evolutionary multiobjective optimization, *IEEE Transactions on Evolutionary Computation*, 12(2), 143–159, 2008.

In our simulation, every variable has a precision of six decimal places. Therefore, the total lengths of the chromosomes in the population for different test functions can be calculated; they are listed in Table C.1. The details of test functions can found in Appendix B.

References

1. Chan, T. M., Man, K. F., Kwong, S., Tang, K. S., A jumping gene paradigm for evolutionary multiobjective optimization, *IEEE Transactions on Evolutionary Computation*, 12(2), 143–159, 2008.
2. Michalewicz, Z., *Genetic Algorithms + Data Structures = Evolution Programs*, Berlin: Springer-Verlag, 1996.

Appendix D: Design of the Fuzzy PID Controller

The structure of the fuzzy proportional-integral-derivative (PID) controller given in Figure 5.5 is similar to the conventional PID controller, but self-tuned control gains are used. These gains are derived from nonlinear functions of the input signals, while its fuzzy control laws are derived from the classical discrete PID controller. In the design, only simple membership functions and minimal fuzzy logic if-then rules are used. Moreover, the normal processes in fuzzy controlled laws (e.g., the fuzzification, control rule execution, and defuzzification steps) are all embedded in the final formulation of the fuzzy control laws, which are in an explicit conventional form. Thus, the resultant controller simply works just like a conventional PID controller.

The stability of this fuzzy PID controller has been analyzed [1–3] and is guaranteed. Its superior performance has also been demonstrated in many simulations and practical examples [1].

Referring to Figure 5.3, the fuzzy PI + D control law can be derived as follows

$$
\begin{aligned}
u_{PID}(nT) &= u_{PI}(nT - T) + K_{uPI}\Delta u_{PI}(nT) \\
&\quad + u_D(nT - T) - K_{uD}\Delta u_D(nT)
\end{aligned}
\tag{D.1}
$$

where $\Delta u_{PI}(nT)$ and $\Delta u_D(nT)$ are the incremental control outputs of the fuzzy PI controller and the fuzzy D controller, respectively; and T is the sampling time. The fuzzy PI and D controllers are then designed by the standard procedure of fuzzy controller design, which consists of fuzzification, control rule base establishment, and defuzzification [4,5]. A fuzzification unit is used to transform the numerical input signal into some fuzzy values, while the defuzzification step is used to transform the final fuzzy value into an output signal from the controller. These two steps require heuristic rules and membership functions to encode the desired system response characteristics and controller dynamics.

In this design, simple membership functions of the input and output of the fuzzy PI and fuzzy D controllers are used, as shown in Figures D.1 and D.2, respectively. Based on a simple control rule and the centroid defuzzification

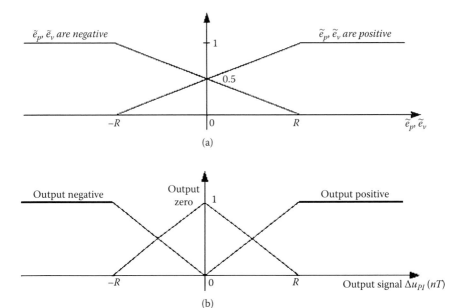

FIGURE D.1
(a) Input membership function and (b) output membership function for the PI component. (From Tang, K. S., Man, K. F., Chen, G., Kwong, S., An optimal fuzzy PID controller, *IEEE Transactions on Industrial Electronics*, 48(4), 757–765, 2001.)

formula, it can be derived that the incremental control output $\Delta u_{PI}(nT)$ is governed by

$$\Delta u_{PI}(nT) = \begin{cases} \dfrac{R[K_i e_p(nT) + K_p e_v(nT)]}{2(2R - K_i |e_p(nT)|)} & \text{in } \Omega_1,\Omega_2,\Omega_5,\Omega_6 \\[2ex] \dfrac{R[K_i e_p(nT) + K_p e_v(nT)]}{2(2R - K_p |e_v(nT)|)} & \text{in } \Omega_3,\Omega_4,\Omega_7,\Omega_8 \\[2ex] \dfrac{1}{2}[K_p e_v(nT) + R] & \text{in } \Omega_9,\Omega_{10} \\[2ex] \dfrac{1}{2}[K_i e_p(nT) + R] & \text{in } \Omega_{11},\Omega_{12} \\[2ex] \dfrac{1}{2}[K_p e_v(nT) - R] & \text{in } \Omega_{13},\Omega_{14} \\[2ex] \dfrac{1}{2}[K_i e_p(nT) - R] & \text{in } \Omega_{15},\Omega_{16} \\[2ex] 0 & \text{in } \Omega_{18},\Omega_{20} \\[1ex] R & \text{in } \Omega_{17} \\[1ex] -R & \text{in } \Omega_{19} \end{cases} \qquad \text{(D.2)}$$

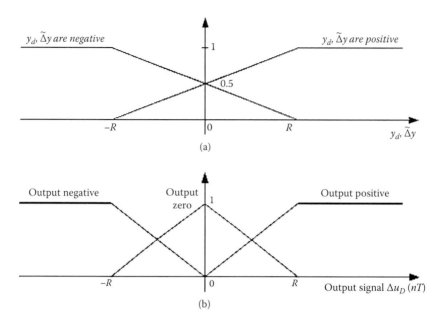

FIGURE D.2
(a) Input membership function and (b) output membership function for the D component. (From Tang, K. S., Man, K. F., Chen, G., Kwong, S., An optimal fuzzy PID controller, *IEEE Transactions on Industrial Electronics*, 48(4), 757–765, 2001.)

where the regions are specified as in Figure D.3a. Similarly, $\Delta u_D (nT)$ can be computed as

$$\Delta u_D (nT)=\begin{cases} \dfrac{R[Ky_d(nT)-K_d\Delta y(nT)]}{2(2R-|y_d(nT)|)} & \text{in } \Omega_1,\Omega_2,\Omega_5,\Omega_6 \\[2ex] \dfrac{R[Ky_d(nT)-K_d\Delta y(nT)]}{2(2R-K_d|\Delta y(nT)|)} & \text{in } \Omega_3,\Omega_4,\Omega_7,\Omega_8 \\[2ex] \dfrac{1}{2}[-K_d\Delta y(nT)+R] & \text{in } \Omega_9,\Omega_{10} \\[2ex] \dfrac{1}{2}[y_d(nT)-R] & \text{in } \Omega_{11},\Omega_{12} \\[2ex] \dfrac{1}{2}[-K_d\Delta y(nT)-R] & \text{in } \Omega_{13},\Omega_{14} \\[2ex] \dfrac{1}{2}[y_d(nT)+R] & \text{in } \Omega_{15},\Omega_{16} \\[2ex] 0 & \text{in } \Omega_{17},\Omega_{19} \\[1ex] -R & \text{in } \Omega_{18} \\[1ex] R & \text{in } \Omega_{20} \end{cases}$$

(D.3)

where the regions are specified as in Figure D.3b.

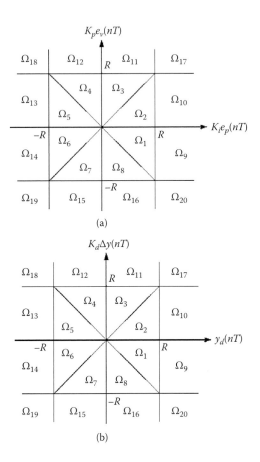

FIGURE D.3

(a) Region for ΔuPI and (b) region for ΔuD of fuzzy PI and D controllers input-combination values. (From Tang, K. S., Man, K. F., Chen, G., Kwong, S., An optimal fuzzy PID controller, *IEEE Transactions on Industrial Electronics*, 48(4), 757–765, 2001.)

It should be noted that a single constant R can be used in these membership functions since the inputs and outputs will be weighted by the gains K_p, K_i, K_d, K_{uPI}, and K_{uD}.

References

1. Carvajal, J., Chen, G., Ogmen, H., Fuzzy PID controller: Design, performance evaluation, and stability analysis, *Information Sciences*, 123, 249–270, 2000.
2. Chen, G., Conventional and fuzzy PID controllers: An overview, *International Journal of Intelligent Control Systems*, 1, 235–246, 1996.

3. Chen, G., Ying, H., BIBO stability of nonlinear fuzzy PI control systems, *International Journal of Intelligent Control Systems*, 5, 3–21, 1997.

4. Misir, D., Malki, H. A., Chen, G., Design and analysis of a fuzzy proportional-integral-derivative controller, *International Journal of Fuzzy Sets and Systems*, 79, 297–314, 1996.

5. Tang, K. S., Man, K. F., Chen, G., Kwong, S., An optimal fuzzy PID controller, *IEEE Transactions on Industrial Electronics*, 48(4), 757–765, 2001.

Index

Printed and bound by CPI Group (UK) Ltd, Croydon, CR0 4YY

18/10/2024

01776236-0003